Introduction To IP Video:

Digitization, Compression and Transmission

Lawrence Harte

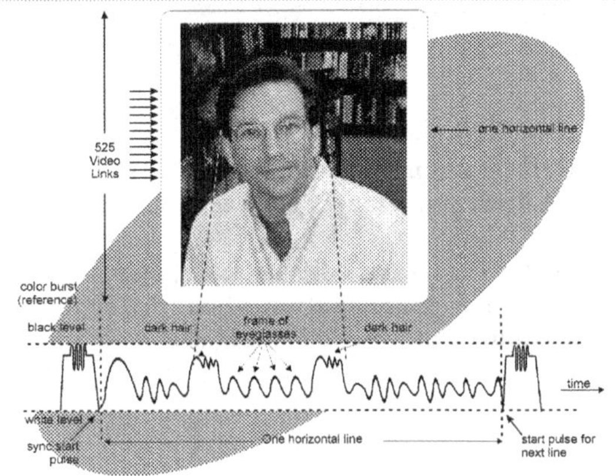

Analog NTSC and PAL Video

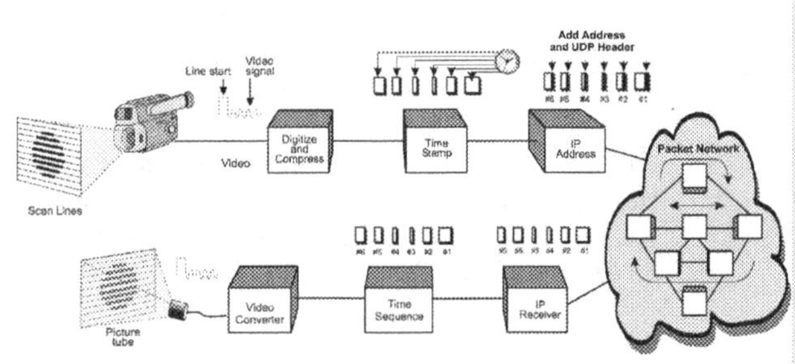

IP Video Transmission

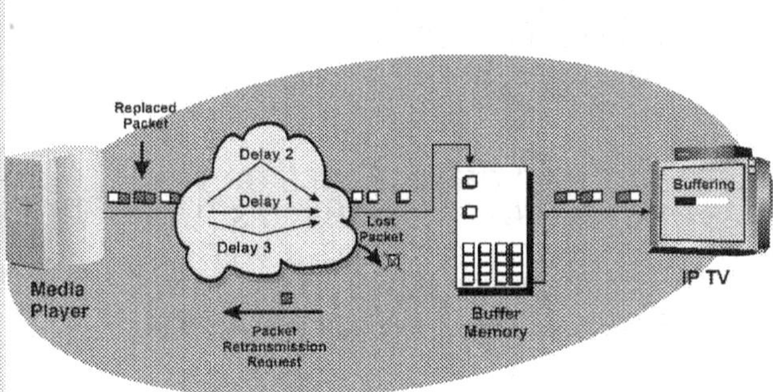

Packet Buffering

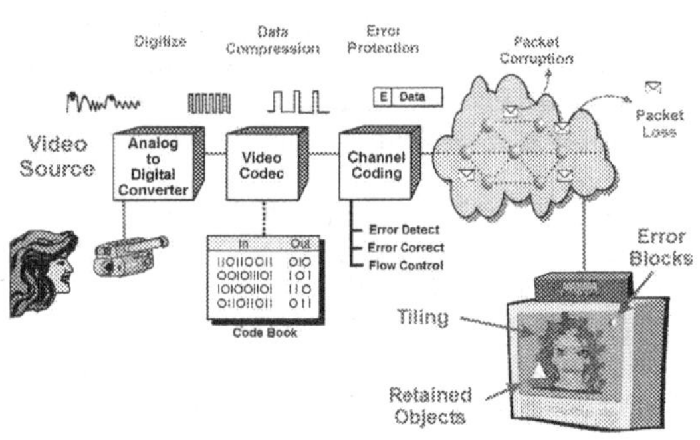

Digital Video Quality

Excerpted From:

IPTV Basics

With Updated Information

ALTHOS Publishing

ALTHOS Publishing

ISBN: 1-932813-41-1

ALTHOS electronic books (ebooks) and images are available for use in educational, promotional materials, training programs, and other uses. For more information about using ALTHOS ebooks and images, please contact April Wiblitzhouser at awiblitzhouser@althos.com or (919) 557-2260

About the Authors

Mr. Harte is the president of Althos, an expert information provider which researches, trains, and publishes on technology and business industries. He has over 29 years of technology analysis, development, implementation, and business management experience. Mr. Harte has worked for leading companies including Ericsson/General Electric, Audiovox/Toshiba and Westinghouse and has consulted for hundreds of other companies. Mr. Harte continually researches, analyzes, and tests new communication technologies, applications, and services. He has authored over 80 books on telecommunications technologies and business systems covering topics such as mobile telephone systems, data communications, voice over data networks, broadband, prepaid services, billing systems, sales, and Internet marketing. Mr. Harte holds many degrees and certificates including an Executive MBA from Wake Forest University (1995) and a BSET from the University of the State of New York, (1990).

Table of Contents

Introduction to IP Video

Introduction to IP Video

IP video is the transfer of video information in IP packet data format. Transmission of IP video involves digitizing video, coding, addressing, transferring, receiving, decoding and converting (rendering) IP video data into its original video form.

Figure 1.1 shows how video can be sent via an IP transmission system. This diagram shows that an IP video system digitizes and reformats the original

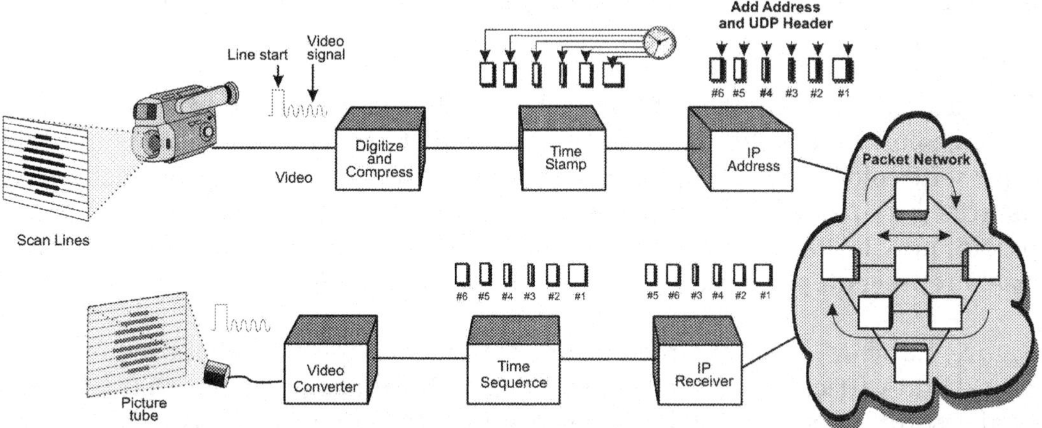

Figure 1.1, IP Video System

video, codes and/or compresses the data, adds IP address information to each packet, transfers the packets through a packet data network, recombines the packets and extracts the digitized video, decodes the data and converts the digital video back into its original video form.

Resolution

Resolution is the number of image elements (pixels) per unit of area. A display with a finer grid contains more pixels, and therefore has a higher resolution capable of reproducing more detail in an image.

Resolution is usually defined as the amount of resolvable detail in the horizontal and vertical directions in a picture. Horizontal resolution usually is expressed as the number of distinct pixels or vertical lines that can be seen in the picture. Vertical resolution is the amount of resolvable detail in the vertical direction in a picture. Vertical resolution usually is expressed as the number of distinct horizontal lines that can be seen.

A pixel is the smallest component in an image. Pixels can range in size and shape and are composed of color (possibly only black on white paper) and intensity. The number of pixels per unit of area is called the resolution and more pixels per unit area provide more detail in the image. Each pixel can be characterized by the amount of intensity and color variations (depth) it can have.

Pixel depth is the number of bits per pixel that is used to represent intensity and color. Pixel color depth is the number of bits per pixel that is used to represent color. For example, a color image that uses 8 bits for each color uses 24 bits per pixel. For a low color depth like 8 bits per pixel, a color palette may be used to map each value to one of a small number of more precisely represented colors, for example to one of 256 24-bit colors.

Figure 1.2 shows how the resolution of a video display is composed of horizontal and vertical resolution components. This diagram shows that the vertical resolution is described as lines per inch (lpi) and the horizontal resolution is described as dots (pixels) per inch (dpi).

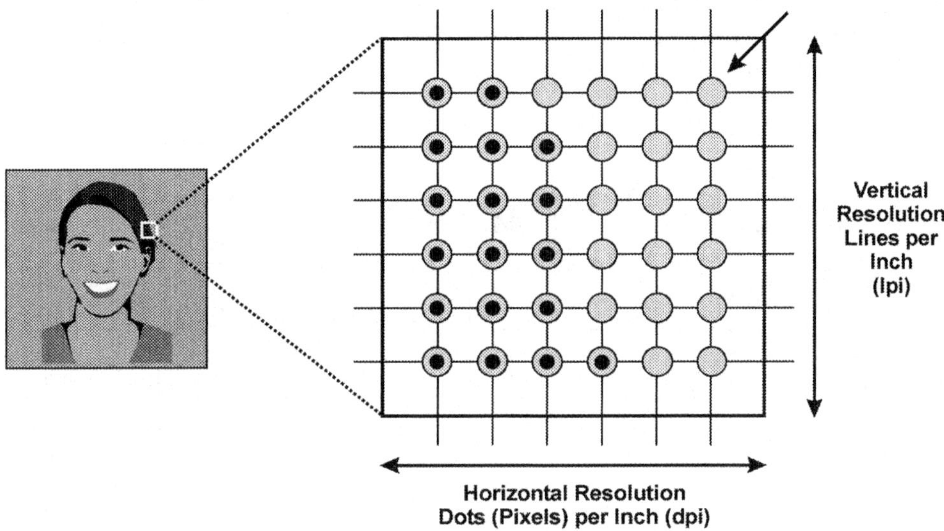

Figure 1.2, Video Resolution

Display formatting is the positioning and timing of graphic elements on a display area (such as on a television or computer display). Television displays can have various resolution and aspect formats. Standard definition (SD) televisions have of 480 lines with 60 interlaced fields (60i) per second for NTSC and 576 lines with 50 interlaced fields (50i) per second for PAL/SECAM. High definition (HD) television is the resolutions of enhanced analog television and digital television. The resolutions of HD range from 480/60p - 480 pixels (vertical) by 728 pixels (horizontal) with 60 progressive fields (60p) per second to 1080/60p - 1080 pixels (vertical) by 1920 pixels (horizontal) with 60 progressive fields per second.

Figure 1.3 shows some of the common types of video display formats for SD and HD televisions. This table shows that standard definition formats can have standard screen aspect ratio 4:3 or widescreen aspect ratio of 16:9. High definition formats have widescreen aspect ratios of 16:9 and have 2 to 4 times higher resolution displays.

	Resolution	Aspect Ratio	Frame Rate
SDTV	704 x 480	16:9	24p, 30p, 60i, 60p
	704 x 480	4:3	24p, 30p, 60i, 60p
	640 x 480	4:3	24p, 30p, 60i, 60p
HDTV			
	1920 x 1080	16:9	24p, 30p, 60i
	1280 x 720	16:9	24p, 30p, 60p

i = Interlaced
p = progressive

Figure 1.3, Display Formats

Frame Rates

Frame rate is the number of images (frames or fields) that are displayed to a movie viewer over a period of time. Frame rate is typically indicated in frames per second (fps).

In general, the higher the frame rate, the better the quality of the video image. When the frame rate is too low, flicker (fluctuations in the brightness of movie images) begins to occur. Flicker typically happens when the frame rate is below 24 frames per second. Some of the common frame rates for moving pictures are 24 fps for film, 25 fps for European video, 30 fps for North American video, 50 fps for European television and 60 fps for North American television.

Figure 1.4 shows the different types of frame rates and how lower frame rates can cause flicker in the viewing of moving pictures. This example shows that frame rates are the number of images that are sent over time (1 second). At the frame rate is reduced below approximately 24 frames per second (fps), the images appear to flicker. Increasing the frame rate results in increased bandwidth requirements. The common frame rate formats are 24 fps for film, 25 fps for European video, 30 fps for North American video, 50 fps for European television and 60 fps for North American television.

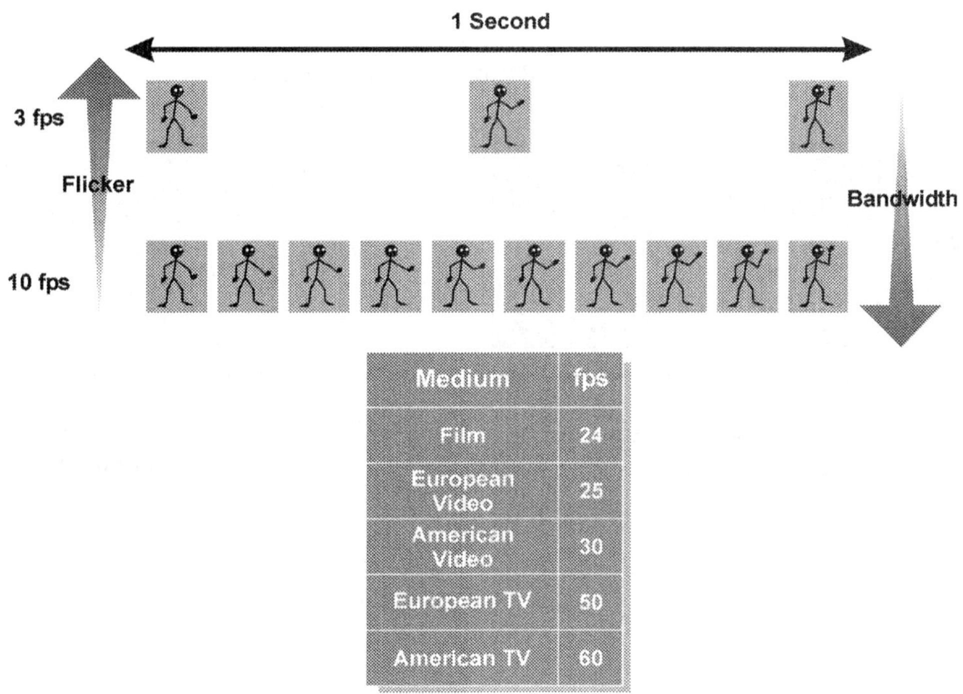

Medium	fps
Film	24
European Video	25
American Video	30
European TV	50
American TV	60

Figure 1.4, Moving Picture Frame Rates

Aspect Ratio

Aspect ratio is the ratio of the number of items (such as pixels on a screen) as compared to their width and height. The aspect ratio determines the frame shape of an image. The aspect ratio of the NTSC (analog television) standard is 4:3 for conventional monitors such as home television sets and 16:9 for HDTV (widescreen).

Figure 1.5 shows how aspect ratio is the relationship between width and height expressed as width:height. This diagram shows that wide screen television has an aspect ratio of 16:9 and that standard television and computer monitors have an aspect ratio of 4:3.

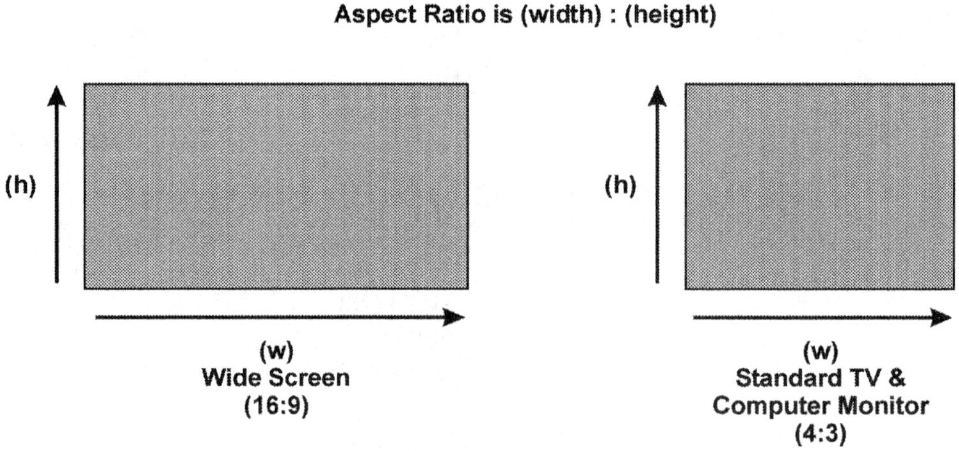

Aspect Ratio is (width) : (height)

(h)
(w)
Wide Screen
(16:9)

(h)
(w)
Standard TV &
Computer Monitor
(4:3)

Figure 1.5, Aspect Ratio

Letterbox

Letterbox is the method of displaying wide screen images on a standard TV receiver where the wide screen aspect ratio is much larger than the standard television or computer monitor typical aspect ratio of 4:3. This causes the display to appear within borders at the top and bottom of the image producing a horizontal box (the letterbox).

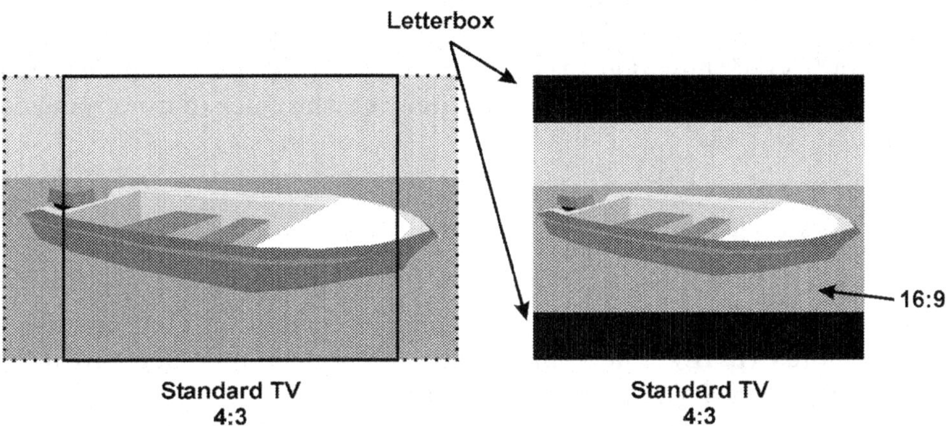

Figure 1.6, Letterbox

Figure 1.6 shows how the use of a letterbox allows an entire video image to be displayed on a screen that has an aspect ratio lower than the video image requires. This example shows that part of an image (of a boat) is lost on the left and right parts of the display. Using a letterbox, the image size is reduced so its width can fit within the length of the screen area. The result is part of the top and bottom areas of the screen area are blanked out (black) resulting in the formation of a box (a letterbox).

Analog Video

Analog video is the representation of a series of multiple images (video) through the use of a rapidly changing signal (analog). This analog signal indicates the luminance and color information within the video signal.

Sending a video picture involves the creation and transfer of a sequence of individual still pictures called frames. Each frame is divided into horizontal and vertical lines. To create a single frame picture on a television set, the frame is drawn line by line. The process of drawing these lines on the screen is called scanning. The frames are drawn to the screen in two separate scans. The first scan draws half of the picture and the second scan draws in between the lines of the first scan. This scanning method is called interlacing. Each line is divided into pixels that are the smallest possible parts of the picture. The number of pixels that can be displayed determines the resolution (quality) of the video signal. The video signal television picture into three parts: the picture brightness (luminance), the color (chrominance) and the audio.

Component Video

Component video consists of three separate primary color signals: red, green, and blue (RGB). The combination of component video can produce any color and intensity of picture information.

Composite Video

Composite video is a single electrical signal that contains luminance, color, and synchronization information. Composite video signals are created by combining several analog signals. Examples of composite video formats include Y, U and V. NTSC, PAL, and SECAM.

The Y intensity (brightness) component includes can be used as a black and white (monochrome) video. The Y component contains the synchronization signal (sync pulse) that coordinates the picture tube scanning (image creation) process. The color components are represented by the two signals U and V. U is the difference between Blue and Y (intensity) and V is the different between Red and Y. The U and V are mixed with different phases (orthogonal) of a color carrier signal and combined to form a chrominance signal.

Figure 1.7 shows how a composite video signal is combines video intensity (Y) and color signals (U and V) to produce a composite video signal. This example shows that an intensity only (black and white) signal is combined with color component signals. The color component signals are added out of phase relative to each other.

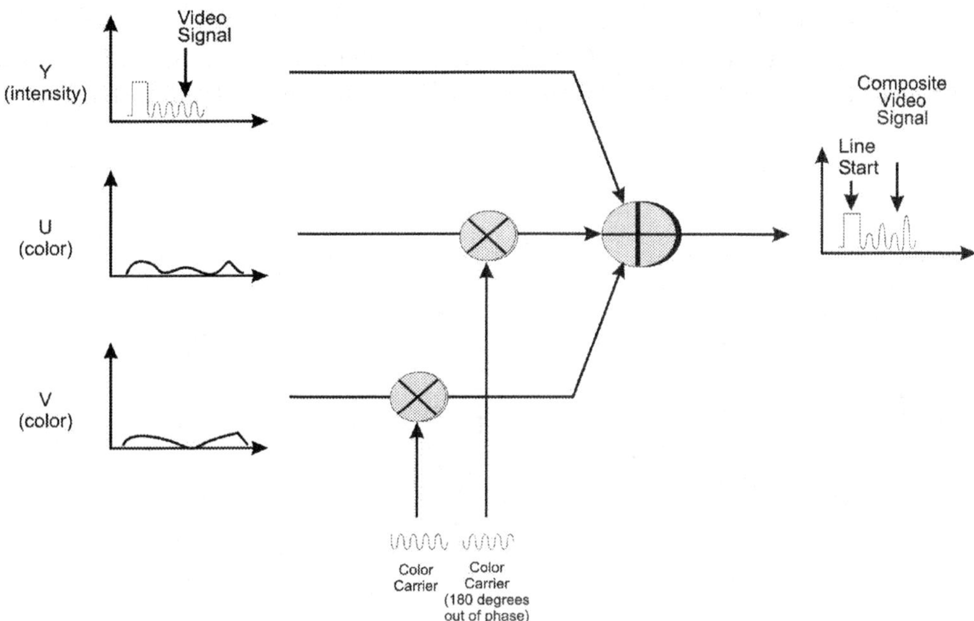

Figure 1.7, Composite Video

Separate Video (S-Video)

S-video is a set of electrical signals that represent luminance and color information. S-video is a type of component video as the intensity and color components are sent as separate signals. This allows the bandwidth of each component video part to be larger than for the signals combined in composite video and it eliminates the cross talk distortion that occurs when multiple signals are combined. This offers the potential for S-video to have improved quality as compared to composite video.

Figure 1.8 shows the pin configurations and signal types used on the S-video connectors. This example shows that the small connector (mini 4 pin) contains a C (chrominance color) video pin #4 and a Y (luminance intensity) video pin #3 and the other two pins are used for a ground (return path) connection. The standard 7 pin S-Video connector contains a C (chrominance color) video pin #4 and a Y (luminance intensity) video pin #3 along with a tuner control data line pin #6 and a tuner clock control line #5 and the remaining two pins (#1 and #2) are used for a ground (return path)

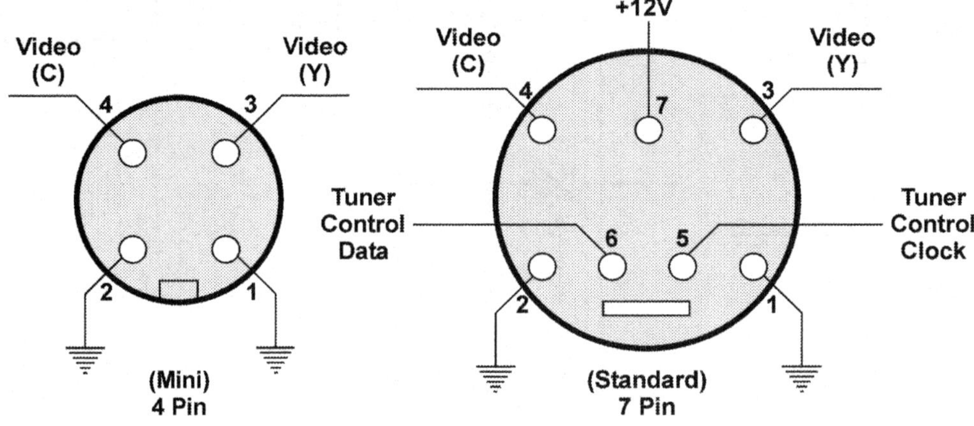

Figure 1.8, S Video Connector Diagrams

Progressive Video

Progressive video is a movie that is composed of individual images where each image contains complete information about a new time period. Progressive frames are symbolized by adding the letter "p" to the frame rate. For example, 30p represents a image rate of 30 progressive frames.

Interlacing

Field interlacing is the process used to create a single video frame by overlapping two pictures where one picture provides the odd lines and the other picture provides the even lines. Field interlacing is performed to reduce bandwidth and flicker.

Figure 1.9 shows examples of progressive and interlaced video. This example shows each image in a progressive video sequence is independent of other images. Images in interleaved video are divided into odd and even lines which are alternated between each image frame. In frame one, every odd line (e.g. 1,3,5, etc) is displayed. In frame two, every even line (e.g. 2,4,6, etc) is displayed. In frame 3, the odd lines are displayed. This process alternates very quickly so the viewer does not notice the interlacing operation.

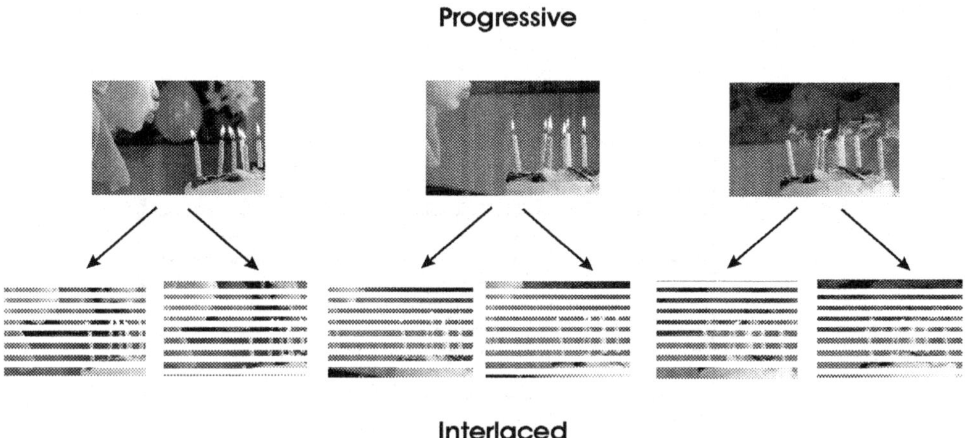

Figure 1.9, Progressive and Interlaced Video

Deinterlacing

Deinterlacing is the process of converting interlaced images into a form that does not contained interlaced images. Deinterlacing is used to convert video

NTSC Video

NTSC video is an established standard for TV transmission that currently in use in the United States, Canada, Japan and other countries. The abbreviation NTSC is often used to describe the analog television standard that transmits 60 fields/seconds (30 frames or pictures/second) and a picture composed of 525 horizontal scan lines, regardless of whether or not a color image is involved.

Figure 1.10 demonstrates the operation of the basic NTSC analog television system. The video source is broken into 30 frames per second and converted into multiple lines per frame. Each video line transmission begins with a burst pulse (called a sync pulse) that is followed by a signal that represents color and intensity. The time relative to the starting sync is the position on the line from left to right. Each line is sent until a frame is complete and the next frame can begin. The television receiver decodes the video signal to position and control the intensity of an electronic beam that scans the phosphorus tube ("picture tube") to recreate the display.

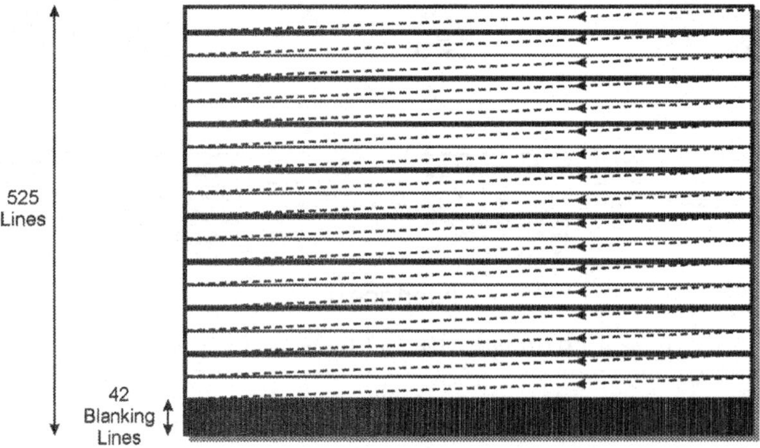

Figure 1.10, Analog NTSC Video

PAL Video

Phase alternating line video is a television system that was developed in the 1980's to provide a common television standard in Europe. PAL is now used in many other parts of the world. The PAL system uses 7 or 8 MHz wide radio channels. The PAL system provides 625 lines of resolution (50 are blanking lines).

Figure 1.11 demonstrates the operation of the basic PAL analog television system. The video source is broken into 30 frames per second and converted into 625 lines per frame. Each video line transmission begins with a burst pulse (called a sync pulse) that is followed by a signal that represents color and intensity. The time relative to the starting sync is the position on the line from left to right. Each line is sent until a frame is complete and the next frame can begin. The television receiver decodes the video signal to position and control the intensity of an electronic beam that scans the phosphorus tube ("picture tube") to recreate the display.

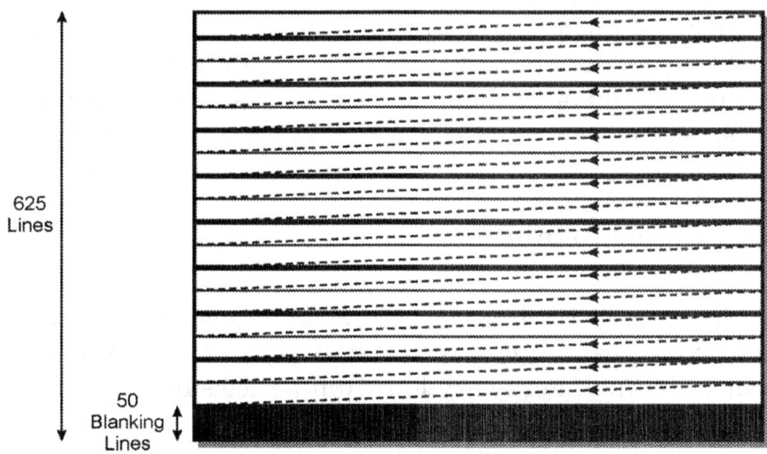

Figure 1.11, Analog PAL Video

SECAM Video

SECAM is an analog color TV system that provides 625 lines per frame and 50 fields per second. This system is similar to PAL and is used in France, the former USSR, the former Eastern Block countries, and some Middle East countries. In order to transmit color the information is transmitted sequentially on alternate lines as a FM signal.

Digital Video

Digital video is a sequence of picture signals (frames) that are represented by binary data (bits) that describe a finite set of color and luminance levels. Sending a digital video picture involves the conversion of an image into digital information that is transferred to a digital video receiver. The digital information contains characteristics of the video signal and the position of the image (bit location) that will be displayed.

Display formatting is the positioning and timing of graphic elements on a display area (such as on a television or computer display). Because the human eye is more sensitive to light intensity than it is to color, display formats can have more intensity components than color components. Some of the common display formats include 4:2:2, 4:2:0, SIF, CIF and QCIF.

4:2:2 Digital Video Format

4:2:2 digital video is a CCIR digital video format specification that defines the ratio of luminance sampling frequency as related to sampling frequencies for each color channel. For every four luminance samples, there are two samples of each color channel.

Figure 1.12 shows the format of a 4:2:2 digital video on a display. In this example, a small portion of the video display has been expanded to show horizontal lines and vertical sample points of luminance (intensity) and chrominance (color). This example shows that the sample frequency for luminance for 4:2:2. is 13.5 MHz and the sample frequency for color is 6.75 MHz. This format has color samples (Cb and Cr) on every line that occur for every other luminance sample.

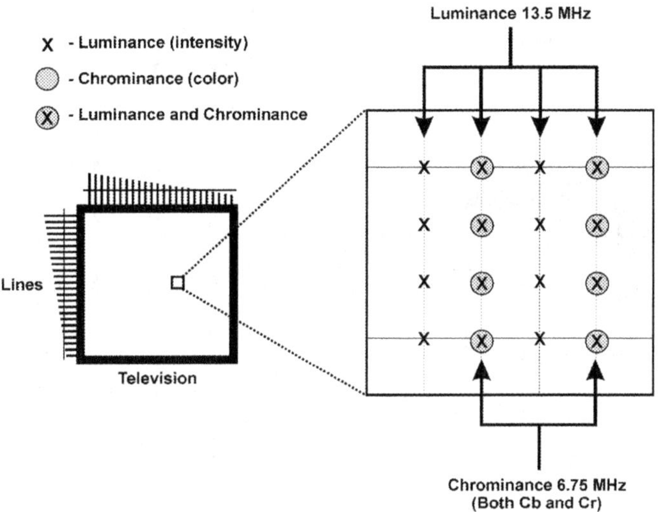

Figure 1.12, 4:2:2 Digital Video Format

4:2:0 Digital Video Format

4:2:0 digital video is a CCIR digital video format specification that defines the ratio of luminance sampling frequency as it is related to sampling frequencies for each color channel. For every four luminance samples, there is one sample of each color channel that alternates on every other horizontal scan line.

Figure 1.13 the format of a 4:2:0 digital video on a display. In this example, a small portion of the video display has been expanded to show horizontal lines and vertical sample points of luminance (intensity) and chrominance (color). This example shows that the sample frequency for luminance for

4:2:0. is 13.5 MHz and the sample frequency for color is 6.75 MHz. This format has color samples (Cb and Cr) on every other line and that the color samples occur for every other luminance sample.

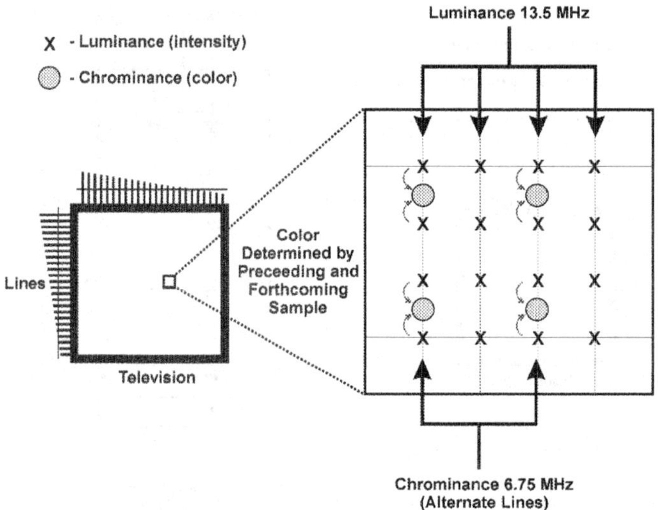

Figure 1.13, 4:2:0 Digital Video Format

Source Intermediate Format (SIF)

Source intermediate format is a digital video format having approximately 1/2 the resolution of analog television (PAL/NTSC). SIF has a luminance resolution of 360 x 288 (625 lines) or 360 x 240 (525 lines) and a chrominance resolution of 180 x 144 (625 lines) or 180 x 120 (525 lines).

Figure 1.14 the format of a SIF digital video on a display. In this example, a portion of the video display 8 x 8 has been expanded to show horizontal lines and vertical sample points of luminance (intensity) and chrominance (color). This example shows that the sample frequency for luminance for SIF is 6.75 MHz and the sample frequency for color is 3.375 MHz. This example shows that the color samples occur on every 4th line and that the color samples occur for every other luminance sample.

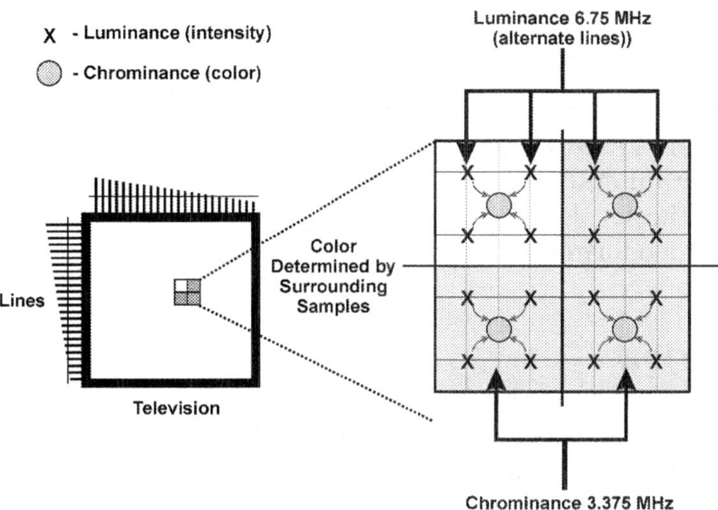

Figure 1.14, SIF Digital Video Format

Common Intermediate Format (CIF)

Common interchange format (CIF) is an image resolution format that is 360 pixels across by 248 pixels high (360x248). The CIF standard is defined in the ITU H.261 and H.264 compression standards. CIF coding includes interframe prediction (using key frames and difference frames), mathematical transform coding and motion compensation.

Quarter Common Intermediate Format (QCIF)

Quarter common interchange format (QCIF) is an image resolution format that is 180 pixels across by 144 pixels high (180x144). The QCIF standard was developed in 1990 is defined by the ITU in as H.261 and H.264 compression standards. H.261 coding includes interframe prediction (using key frames and difference frames), mathematical transform coding and motion compensation.

Video Digitization

Video digitization is the conversion of video component signals or composite signal into digital form through the use of an analog-to-digital (pronounced A to D) converter. The A/D converter periodically senses (samples) the level of the analog signal and creates a binary number or series of digital pulses that represent the level of the optical image. Digital video may be created from another video format (e.g. analog video) or from film.

Video Capturing

Video capturing is the process of receiving and storing video images. Video capture typically refers to capture of and conversion (quantization) of video images into digital form.

Quantization

Quantization is the process of representing sample signals with a range if defined values. In analog-to-digital conversion, a continuous analog value is represented by one of a finite number of quantized values. In lossy signal compression (such as an A-law encoding) one digital value is represented by another one which is usually not precisely the same. Except in lucky cases where the quantized value is exactly the same as the original, quantization introduces error (or noise).

Quantization noise (or distortion) is the error that results from the conversion of a continuous analog signal into a finite number of digital samples that can not accurately reflect every possible analog signal level. Quantization noise is reduced by increasing the number of samples or the number of bits that represent each sample. This term also is known as quantization distortion.

Figure 1.15 shows a fundamental process that can be used to digitize moving pictures or analog video into digital video. For color images, the image is filtered into red, green and blue component colors. Each of the resulting images is scanned in lines from top to bottom to convert the optical level to an equivalent electrical level. The electrical signal level is periodically sampled and converted to its digital equivalent level. This example shows that analog signals can have 256 levels (0-255) and that this can be represented by 8 bits of information (a byte).

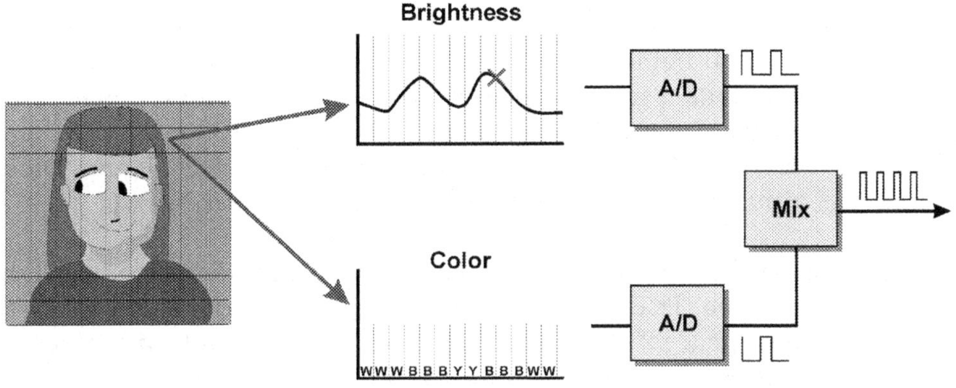

Figure 1.16, Video Digitization

Analog video is typically stored and transmitted in interlaced format where a single frame is converted into two interlaced fields. Analog video is converted into a progressive form by deinterlacing. Deinterlacing is the process of converting interlaced images into a form that of unique sequential images.

Digital video is usually stored in progressive format where each image in a sequence of a movie adds a new image (as opposed to a partial interlaced image) progression in a moving picture. When video is stored in progressive form, it is identified by the letter p at the end of the frame rate (e.g. 60p is 60 images per second and each frame is unique).

Film to Video Conversion

Film to video conversion is the process of scanning film (e.g. movie) images and converting it into a sequence of video images. Because the frame rate of film is different than video, the frame rate is adapted through the use of a pulldown process.

3:2 pulldown is a process of converting film that operates at 24 frames per second to video at 60 fields per second. The 3:2 process operates by repeating a film frame image 3 times, repeating the next film frame image 2 times and repeating this process 3 repeats and 2 repeats to create 60 images (fields) per second. 3:2 pulldown is sometimes called Telecine because Telecine was a machine that performed the pulldown conversion.

Figure 1.17 shows how 24 frame per second (fps) film is converted to 60 field per second (fps) video. To create the 60 frames per second, each frame image must be copied (repeated) 2.5 times. This example shows that frames with in the film are used to create 2 or 3 interlaced video fields.

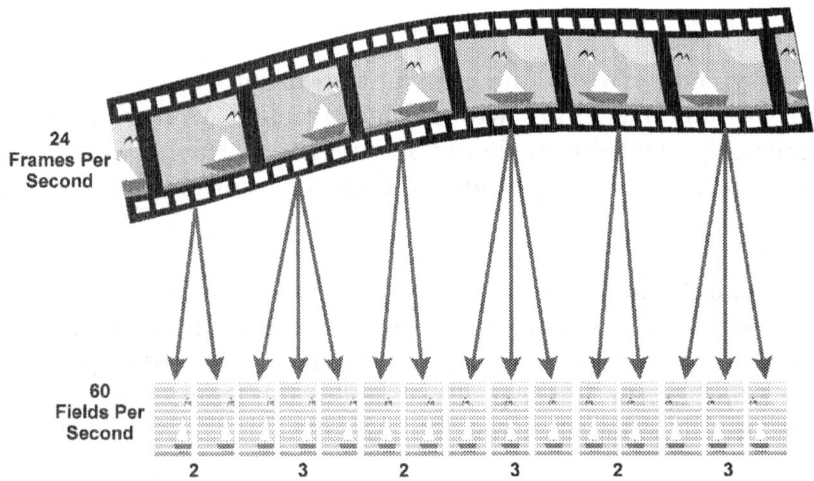

Figure 1.17, Pulldown

Video Compression

Video compression is the process of reducing the amount of transmission bandwidth or the data transmission rate by analog processing and/or digital coding techniques. Moving pictures can be compressed by removing redundancy within each image (spatial redundancy) or between successive images over a period of time (temporal redundancy). When compressed, a video signal can be transmitted on circuits with relatively narrow channel bandwidth or using data rates 50 to 200 times lower than their original uncompressed form.

Spatial Compression (Image Compression)

Spatial compression is the analysis and compression of information or data within a single frame, image or section of information.

One of the common forms of spatial compression is specified by the joint picture experts group (JPEG). JPEG is a working committee under the auspices of the International Standards Organization (ISO) with the goal of defining a standard for digital compression and decompression of still images for use in computer systems. The JPEG committee has produced an image compression standard format that is able to reduce the bit per pixel ratio to approximately 0.25 bits per pixel for fair quality to 2.5 bits per pixel for high quality.

JPEG uses lossy compression methods that result in some loss of the original data. When you decompress the original image, you don't get exactly the same image that you started with despite the fact JPEG was specifically designed to discard information not easily detected by the human eye.

The JPEG committee has defined a set of compression methods that are used to provide for high-quality images at varying levels of compression up to approximately 50:1. The JPEG compression system can use compression that is fully reversible (no loss of information) or that is lossy (reversible with some loss of quality).

Lossy compression is a process of reducing an amount of information (usually in digital form) by converting it into another format that represents the initial form of information. Lossy compression does not have the ability to guarantee the exact recreation of the original signal when it is expanded back from its compressed form.

JPEG compression typically works better for photographs and reference video frames (I frames) rather than line art of cartoon graphics. This is because the compression methods tend to approximate portions of the image and the approximation of lines or sharp boundaries tends to get blurry with unwanted artifacts.

The JPEG compression process begins by dividing digital image into groups of blocks. These blocks are then converted from a pixel domain (bit maps) into a frequency domain (a group of images with different detail levels) using a discrete cosine transform process. These frequency components are then converted into specific levels. The compression system may choose to remove frequency components that have a limited amount of information (low levels) through a threshold process. The data is then compressed using run length encoding top remove (to represent) long sequences by shorter codes (run length encoding) and then by variable length coding to convert repeated sequences over varying lengths into shorter codes (variable length encoding).

Discrete Cosine Transform (DCT)

Discrete cosine transform, is a form of frequency analysis that is applied to discrete signals (e.g. binary data) to produce an output that is composed of the frequency components and the levels (coefficients) that represent the original digital signal. A DCT output is composed of a DC component (basic intensity) and a series of increasing frequency components that reflect the complexity of the underlying data.

Thresholding

Thresholding is the process of modifying numbers or measurements that are within a range or meet some criteria to produce a lower number or a lesser number of data elements. Thresholding is used in lossy data compression processes (such as image compression) to reduce the amount of data through the loss of accuracy of information that has little impact on the user.

Run Length Encoding (RLE)

Run length encoding is a method of compressing digital information by representing repetitive data information by a notation that indicates the data that will be repeated and how many times the data will be repeated (run length).

Variable Length Encoding (VLE)

Variable length encoding is a method of compressing digital information by representing repetitive groups of data information by code that are used to look up the data sequence along with how many times the data will be repeated (variable length).

Figure 1.18 shows the basic process that can be used for JPEG image compression. This diagram shows that JPEG compression takes a portion (block) of a digital image (lines and column sample points) and analyzes the block of digital information into a new block sequence of frequency components (DCT). The sum of these DCT coefficient components can be processed and added together to reproduce the original block. Optionally, the coefficient levels can be changed a small amount (lossy compression) without sig-

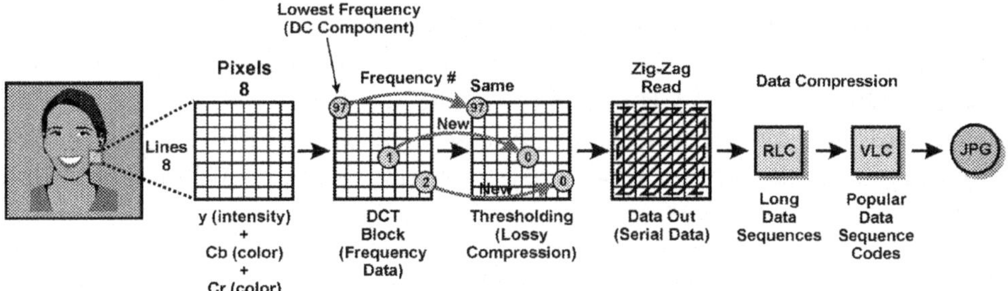

Figure 1.18, JPEG Image Compression

nificant image differences (thresholding). The new block of coefficients is converted to a sequence of data (serial format) by a zig-zag process. The data is then further compressed using run length coding (RLC) to reduce repetitive bit patterns and then using variable length coding (VLC) to convert and reduce highly repetitive data sequences.

Time Compression (Temporal Compression)

Temporal compression is the analysis and compression of information or data over a sequence of frames, images or sections of information.

One of the more common forms of temporal compression used for digital is specified by the Motion Picture Experts Group (MPEG). MPEG is a working committee that defines and develops industry standards for digital video systems. These standards specify the data compression and decompression processes and how they are delivered on digital broadcast systems. MPEG is part of International Standards Organization (ISO).

Temporal video compression involves analyzing the changes that occur between successive images in a video sequence that that only the difference between the images is sent instead of all of the information in each image. To accomplish this, time compression can use keyframes and motion estimation.

Figure 1.19 shows how key frames can be used to reduce the amount of data that is transmitted for video signals. This example shows two scenes in a video clip. The first scene is of a sail boat that is slowly moving across the horizon on the water and the second scene is of a house on the shoreline. This example shows that a key frame is sent at the beginning of a scene and only changes to the key frame are necessary to send. When a new scene occurs, a new key frame is sent.

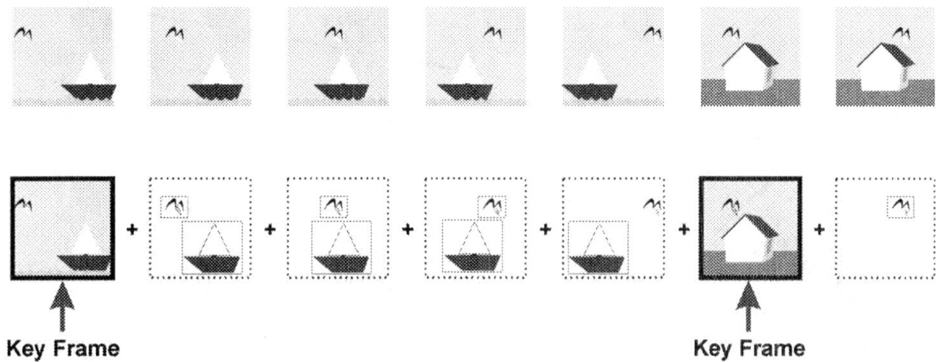

Figure 1.19, Using Key Frames in Digital Video

Motion estimation is the process of searching a fixed region of a previous frame of video to find a matching block of pixels of the same size under consideration in the current frame. The process involves an exhaustive search for many blocks surrounding the current block from the previous frame. Motion estimation is a computer-intensive process that is used to achieve high compression ratios. Block matching is the process of matching the images in a block (a portion of an image) to locations in other frames of a digital picture sequence (e.g. digital video).

Figure 1.20 shows how a digital video system can use motion estimation to identify objects and how their positions change in a series of pictures. This diagram shows that a bird in a picture is flying across the picture. In each picture frame, the motion estimation system looks for blocks that approximate other blocks in previous pictures. Over time, the digital video motion estimation system finds matches and determines the paths (motion vectors) that these objects take.

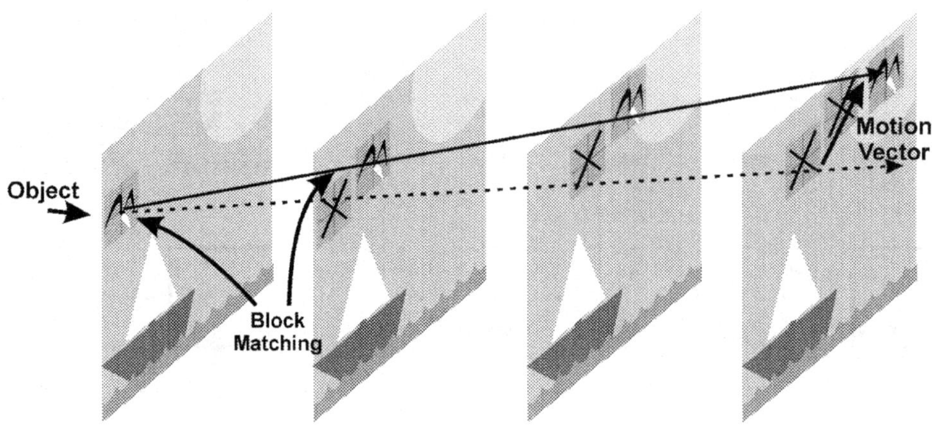

Figure 1.20, Motion Estimation

Coding Redundancy (Data Compression)

Coding redundancy is the repetition of information or bits of data within a sequence of data. Using data compression can reduce coding redundancy. Data compression is a technique for encoding information so that fewer data bits of information are required to represent a given amount of data. Some of the common forms of data compression used in video compression include run length encoding (RLE) and variable length encoding (VLE).

Figure 1.21 shows how video compression may use spatial and temporal compression to reduce the amount of data to represent a video sequence. This diagram shows that a frame in a video sequence may use spatial compression by representing the graphic elements within the frame by objects or codes. The first frame of this example shows that a picture of a bird that is flying in the sky can be compressed by separating the bird image from the blue background and making the bird an object and representing the blue background as a box (spatial compression). The next sequence of images only needs to move the bird on the background (temporal compression).

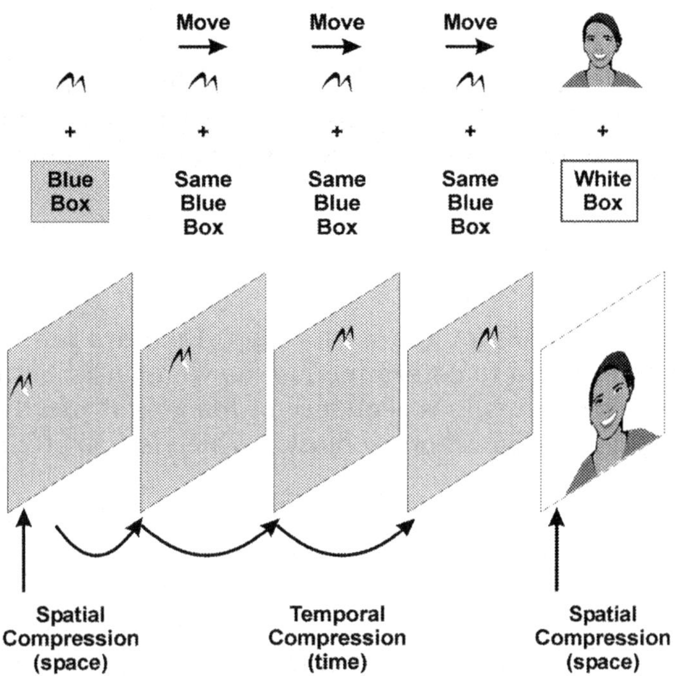

Figure 1.21, Video Compression

Video images are composed of pixels. MPEG system groups pixels within each image into small blocks and these blocks are grouped into macroblocks. Macroblocks can be combined into slices and each image may contain several slices. Slices make up frames, which come in several different types. The different types of frames can be combined into a group of pictures.

Pixels

A pixel is the smallest component in an image. Pixels can range in size and shape and are composed of color (possibly only black on white paper) and intensity. The number of pixels per unit of area is called the resolution. The more pixels per unit area provide more detail in the image.

Blocks

Blocks are portions of an image within a frame of video usually defined by a number of horizontal and vertical pixels. For the MPEG system, each block is composed of 8 by 8 pixels and each block is processed separately.

Macroblocks

A macroblock is a region of a picture in a digital picture sequence (motion pictures) that may be used to determine motion compensation from a reference frame to other pictures in a sequence of images. Typically, a frame is divided into 16 by 16 pixel sized macroblocks, which is also groupings of four 8 by 8-pixel blocks.

Slice

A slice is a part of an image that is used in digital video and it is composed of a contiguous group of macroblocks. Slices can vary in size and shape.

Frames

A frame is a single still image within the sequence of images that comprise the video. In an interlaced scanning video system, a frame comprises two fields. Each field contains half of the video scan lines that make up the picture, the first field typically containing the odd numbered scan lines and the second field typically containing the even numbered scan lines.

To compress video signals, the MPEG system categorizes video images (frames) into different formats. These formats vary from fame types that only use spatial compression (independently compressed) to frames that use both spatial compression and temporal compression (predicted frames).

MPEG system frame types include independent reference frames (I-frames), predicted frames that are based on previous reference frames (P-frames), bidirectionally predicted frames using preceding frames and frames that follow (B-Frames), and DC frames (basic block reference levels).

Intra Frames (I-Frames)

Intra frames (I-Frames) are complete images (pictures) within a sequence of images (such as in a video sequence). I-frames are used as a reference for other compressed image frames and I frames are completely independent of other frames. The only redundancy that can be removed from I frames is spatial redundancy. This means that I-frames require more data than compressed frames.

Predicted Frames (P-Frames)

Predicted frames (P-Frames) are images (pictures) within a sequence of images (such as in a video sequence) that are created using information from other images (such as from I-Frames).

Because image components are often repeated within a sequence of images (temporal redundancy), the use of P-Frames provides substantial reduction in the number of bits that are used to represent a digital video sequence (temporal data compression).

Bi-Directional Frames (B-Frames)

Bi-directional frames (B-Frames) are images (pictures) within a sequence of images (such as in a video sequence) that are created using information from preceding images and images that follow (such as from I-Frames and predicted frames P-Frames).

Because B-Frames are created using both preceding images and images that follow, B-frames offer more data compression capability than P-Frames. B-frames require the use of frames that both precede and follow the B-frames. Because B-frames must be compared to two other frames, the amount of image processing that is required for B-frames (e.g. motion estimation) is typically higher than P frames.

DC Frames (D-Frames)

A DC frame is an image in a motion video sequence that represents the DC level of the image. D frames are used in the MPEG-1 system to allow rapid viewing (e.g. fast forwarding) and are not used in other versions of MPEG.

Groups of Pictures (GOP)

Frames can be grouped into sequences called a group of pictures (GOP). A GOP is an encoding of a sequence of frames that contain all the information that can be completely decoded within that GOP. For all frames within a GOP that reference other frames (such as B-frames and P-frames), the frames so referenced (I-frames and P-frames) are also included within that same GOP.

The types of frames and their location within a GOP can be defined in time (temporal) sequence. The temporal distance of images is the time or number of images between specific types of images in a digital video. M is the distance between successive P-Frames and N is the distance between successive I-Frames. Typical values for MPEG GOP are M equals 3 and N equals 12.

Figure 1.22 shows how different types of frames can compose a group of pictures (GOP). A GOP can be characterized as the depth of compressed predicted frames (m) as compared to the total number of frames (n). This example shows that a GOP starts within an intra-frame (I-frame) and that intra-frames typically require the largest number of bytes to represent the image (200 kB in this example). The depth m represents the number of frames that exist between the I-frames and P-frames.

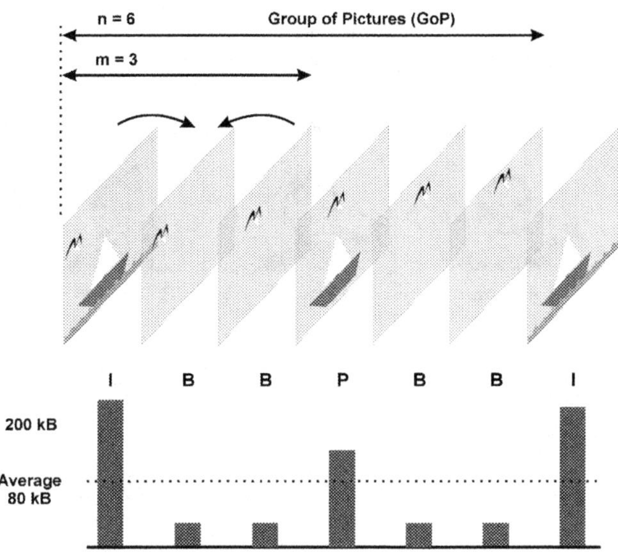

Figure 1.22, MPEG Group of Pictures (GOP)

Groups of pictures can be independent (closed) GOP or they can be relative (open) to other GOPs. An open group of pictures is a sequence of image frames that requires information from other GOPs to successfully decode all the frames within its sequence. A closed group of pictures is a sequence of image frames can successfully decode all the frames within its sequence without using information from other GOPs.

Because P and B frames are created using other frames, when errors occur on previous frames, the error may propagate through additional frames (error retention). To overcome the challenge of error propagation, I frames are sent periodically to refresh the images and remove and existing error blocks.

Figure 1.23 shows how errors that occur in an MPEG image may be retained in frames that follow. This example shows how errors in a B-Frame are transferred to frames that follow as the B-Frame images are created from preceding images.

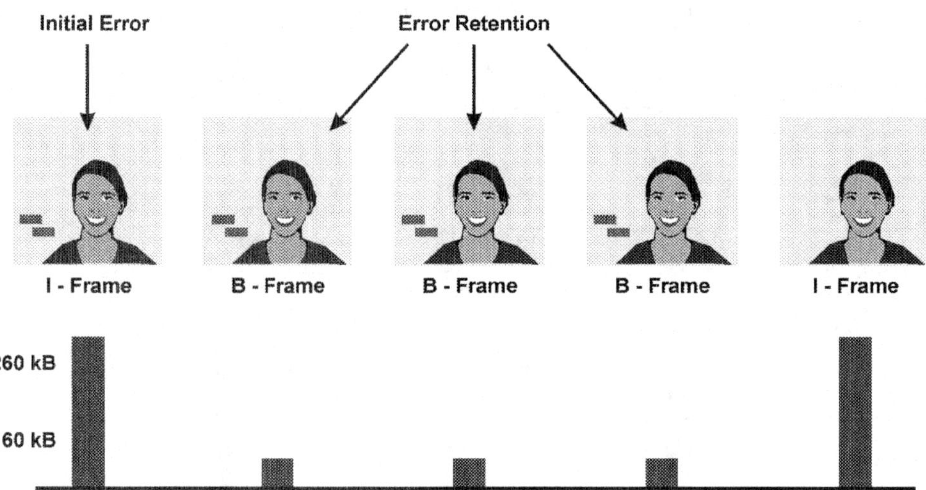

Figure 1.23, MPEG Error Retention

Compression Scalability

Compression scalability is the ability of a media compression system to adapt its compression parameters for various conditions such as display size (spatial scalability), combining multiple transmission channels (layered scalability), changing the frame rate (temporal scalability) or quality of signal (signal to noise scalability).

Spatial scalability is the ability of a media file or picture image to reduce or vary the number of image components or data elements representing a picture over a given area (spatial area) without significantly changing the quality or resolution of the image.

Layered scalability is the use of multiple layers in an image that can be combined to produce higher resolution images or video. Layered compression starts with the use of a base layer may be decoded separately to provide a low resolution preview of the image or video and to reduce the decoding processing requirements (reduced complexity). An enhancement layer is a stream or source of media information that is used to improve (enhance) the resolution or appearance of underlying image (e.g. base layers).

Temporal scalability is the ability of a streaming media program or moving picture file to reduce or vary the number of images or data elements representing that media file for a particular time period (temporal segment) without significantly changing the quality or resolution of the media over time.

Signal to noise ratio scalability is the ability of a media file or picture image to reduce or vary the number of image components or data elements representing that that picture to compensate for changes in the signal to noise ratio of the transport signal.

Layered Video Coding

Layered video coding is a process that converts video into several component parts where each layer can be combined with other layers to produce a higher quality or improved version of the media.

An example of layered video coding is digital video broadcasting (DVB) signals that provide a high bit rate layer (DVB-T) for standard televisions and a low bit rate layer (DVB-H) for handheld devices.

Figure 1.24 shows how layered video coding can be used to provide variable levels of quality that is determined by the number of layers that can be combined. This example shows the base layer provides a rough image of a circle. As additional layers are added, each layer ads further detail (smaller blocks) to the circle.

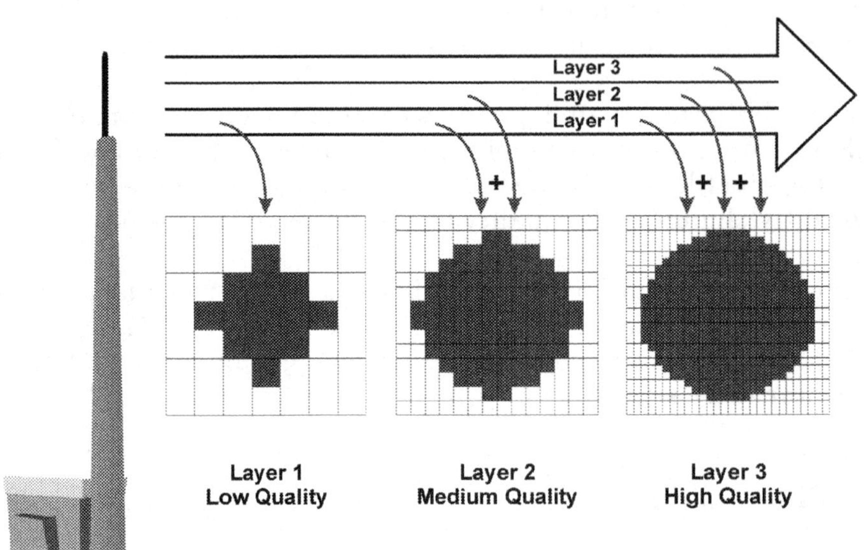

Figure 1.24, Layered Video Coding

Advanced Video Coding (AVC/H.264)

Advanced video coding is a video codec that can be used in the MPEG-4 standard. The AVC coder provides standard definition (SD) quality at approximately 2 Mbps and high definition (HD) quality at approximately 6-8 Mbps.

The AVC coding system achieves higher compression ratios by better analysis and compression of the underlying media. The AVC system can identify and separately code objects from video sequences (object coding), it can create or represent objects in synthetic form (animated objects) and it can use variable block sizes to more efficiently represent (deblock) images with varying edges.

Object Coding

Object coding is the representation of objects (such as a graphic item in a frame of video) through the use of a code or character sequence. The types of objects that can be used in AVC system range from static background images (sprites) to synthetic video (animation).

A background sprite is a graphic object that is located behind foreground objects. Background sprites usually don't change or they change relatively slowly. Audiovisual objects are parts of media images (media elements). Media images or moving pictures may be analyzed and divided into audiovisual objects to allow for improved media compression or audiovisual objects may be combined to form new images or media programs (synthetic video).

Media elements are component parts of media images or content programs. A media element can be the smallest common denominator of an image or media program component. A media element is considered a unique specific element such as a shape, texture and size.

Animated Objects

Animated objects are graphic elements that can be created and changed over a period of time. Animated objects can be used to create synthetic video (e.g. moving picture information that is created through the use creating image components by non-photographic means).

MPEG-4 uses an efficient form (a binary form) of virtual reality modeling language (VRML) to create 3 dimensional images. This allows the MPEG system to send a mathematical model of objects along with their associated textures instead of sending detailed images that require large data transmission bandwidths.

Variable Block Sizes

Variable block sizes are groups of image bits that make up a portion of an image that vary in width and height. The use of variable block sizes allows for using smaller blocks in portions of graphic images that have lots of rapid variations (such as the edge of a sharp curve). Also, it allows larger blocks in areas that have a limited amount of variation (such as the solid blue portion of the sky).

Real Time Encoding

Real time encoding is the process of converting or processing media from one formation or data structure to another format or data structure where the process occurs immediately or within a very short time period after the media is available for conversion.

Non-Real Time Encoding

Non-real time encoding is a process of converting or processing media from one formation or data structure to another format where the amount of time to perform the conversion can exceed the time that would be perceived as real time.

Figure 1.25 shows how two-pass digital video encoding can be used to reduce the data transmission rate for digital video signals. This example shows a scene of a sail boat that is slowly moving across the horizon and a bird that is flying toward the sun. During the first encoding pass, the objects in each image are identified. During the 2nd encoding pass, the change in position between the objects in is identified. By sending only the changed position information, this dramatically reduces the amount of data that represents the digital video signal.

IP Video Transmission

IP video transmission is the transport of video (multiple images) that is in the form of data packets to a receiver through an IP data network. IP video can be downloaded completely before playing (file downloading), transferred as the video is played (streaming) or played as soon as enough of the media has been transferred (progressive downloading).

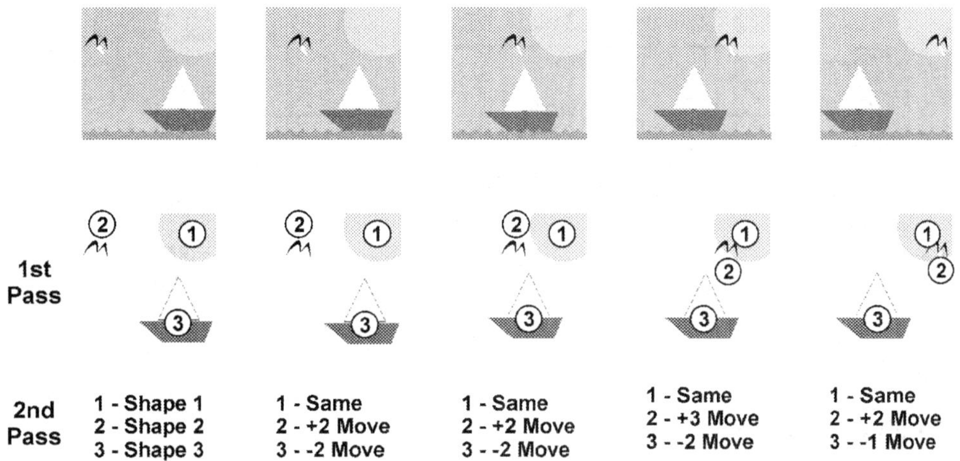

Figure 1.25, Two Pass Video Compression

File Downloading

File download is the transfer of a program or of data from computer server to another computer. File download commonly refers to retrieving files from a web site server to another computer.

Figure 1.26 shows how to download movies through the Internet. This diagram shows how the web server must transfer the entire media file to the media player before viewing can begin.

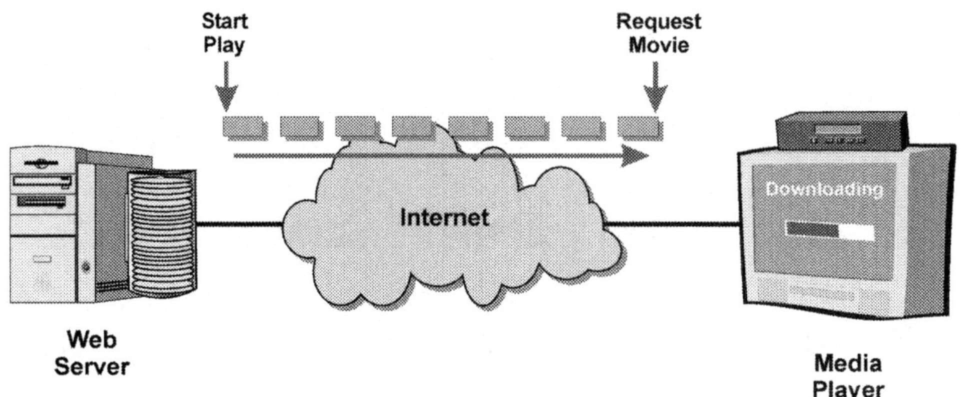

Figure 1.26, Digital Video File Downloading

Video Streaming

Video streaming is the process of delivering video, usually along with synchronized accompanying audio in real time (no delays) or near real time (very short delays). Upon request, a video media server system will deliver a stream of video and audio (both can be compressed) to a client. The client will receive the data stream and (after a short buffering delay) decode the video and audio and play them in synchronization to a user.

Video streaming can be transmitted with or without flow control. When a video streaming session can obtain and use feedback from the receiver, it is called intelligent streaming. Intelligent streaming is the providing of a continuous stream of information such as audio and video content with the ability to dynamically change the characteristics of the streaming media to compensate for changes in the signal source, transmission or media application playing capabilities.

Figure 1.27 shows how to stream movies through the IP data networks. This diagram shows that streaming allows the media player to start displaying the video before the entire contents of the file have been transferred. This diagram also shows that the streaming process usually has some form of feedback that allows the viewer to control the streaming session and to provide feedback to the media server about the quality of the connection. This allows the media server to take action (such as increase or decrease compression and data transmission rate) if the connection is degraded or improved.

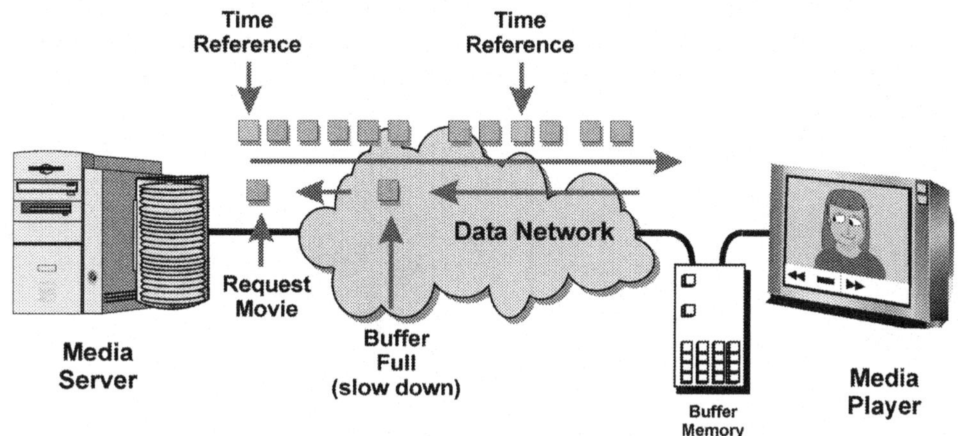

Figure 1.27, Video Streaming

Bandwidth Awareness

Bandwidth awareness is the process and/or protocols that are used to allow for the recognition of the bandwidth that is available to devices that are connected to a network. For example, a media server needs to discover the connection speed of a multimedia computer that has requested to view a digital video stream. If the server has bandwidth awareness capability, it can determine an optimum compression and data transmission rate for the requested digital video stream.

Intelligent Streaming

Intelligent streaming is the providing of a continuous stream of information such as audio and video content with the ability to dynamically change the characteristics of the streaming media to compensate for changes in the signal source, transmission or media application playing capabilities.

Figure 1.28 shows how a media server can adjust its data transmission rate to compensate for different user data rates. This example shows how a media server is streaming packets to an end user (for an Internet television viewer). Some of the packets are lost at the receiving end of the connection because of the access device. The receiving device (a television set top box) sends back control packets to the media server indicating that the communication session is experiencing a higher than desirable packet or frame loss rate. The media server can use this information to change its media compression and data transmission rates to compensate for the slow user access link.

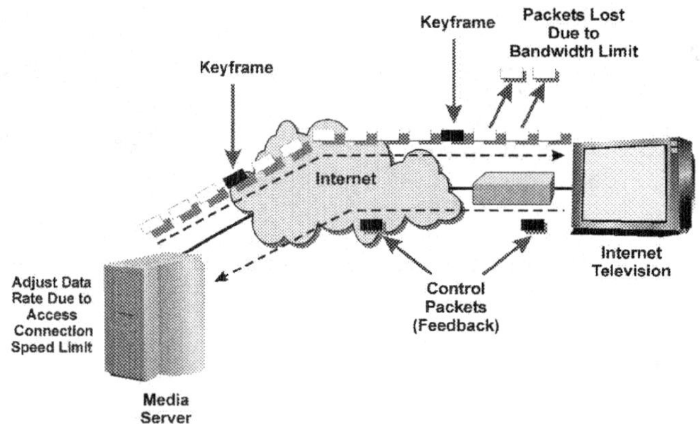

Figure 1.28, Digital Video Streaming Bandwidth Control

Bit Stream Syntax

Bit stream syntax is a set of commands and formats that are used by a sequence of encoded bits that are flowing over a digital communications channel.

Multiple Bit Rate

Multiple bit rate is the storing and/or streaming of media using different bit rates to represent the media (such as low or high resolution digital video formats).

Frame Dropping

Frame dropping is the process of discarding or not using all the video frames in a sequence of frames. Frame dropping may be used to temporarily reduce the data transmission speed or to reduce the video and image processing requirements.

Stream Thinning

Stream thinning is the process of removing some of the information in a media stream (such as removing image frames) to reduce the data transmission rate. Stream thinning may be used to reduce the quality of a media stream as an alternative to disconnecting the communication session due to bandwidth limitations.

Progressive Downloading

Progressive downloading is transferring of a file or data in a sequential process that allows for the using of portions of the data before the transfer is complete. An example of progressive downloading is HTTP streaming. HTTP streaming manages the sequential transferring of media files through an IP data network (such as the Internet) using HTTP commands.

Internet Protocol Packet Transmission

Packet transmission is the transfer of data from one point to another point by dividing (disassembling) the data into short increments, or packets, each of which can be routed separately from a source then reassembled in the proper order at the destination. Some of the key issues for transferring IP Video the packet data networks include transmission delay, packet jitter and packet loss.

Transmission delay is the time that is required for transmission of a signal or packet of data from entry into a transmission system (e.g. transmission line or network) to its exit of the system. Common causes of transmission delay include transmission time through a transmission line (less than the speed of light), channel coding delays, switching delays, queuing delays waiting for available transmission channel time slots, and channel decoding delays. Packet jitter is the undesirable random changes in the arrival rate of packets. Packet loss is a ratio of the number of data packets that have been lost in transmission compared to the total number of packets that have been transmitted.

Figure 1.29 shows how blocks of data are divided into small packet sizes that can be sent through the Internet. After the data is divided into packets (envelopes shown in this example), a destination address along with some description about the contents is added to each packet (called the packet header). As the packet enters into the Internet (routing boxes shown in this diagram), each router reviews the destination address in its routing table and determines which paths it can send the packet to so it will move further towards its destination. If a current path is busy or unavailable (such as shown for packet #3), the router can forward the packets to other routers that can forward the packet towards its destination. This example shows that because some packets will travel through different paths, packets may

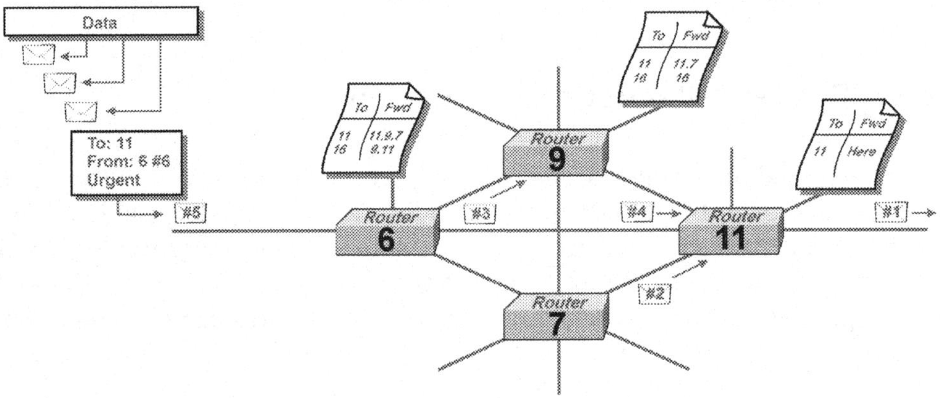

Figure 1.29, IP Packet Transmission

arrive out of sequence at their destination. When the packets arrive at their destination, they can be reassembled into proper order using the packet sequence number.

User Datagram Protocol (UDP)

UDP is a high-level communication protocol that coordinates the one-way transmission of data in a packet data network. The UDP protocol coordinates the division of files or blocks of data information into packets and adds sequence information to the packets that are transmitted during a communication session using Internet protocol (IP) addressing. This allows the receiving end to receive and re-sequence the packets to recreate the original data file or block of data that was transmitted. UDP adds a small amount of overhead (control data) to each packet relative to other high-level protocols such as TCP. However, UDP does not provide any guarantees to data delivery through the network. UDP protocol is defined in request for comments 768 (RFC 768).

Figure 1.30 shows how user datagram protocol (UDP) operates to efficiently send data through a packet network. This diagram shows that the UDP system first packetizes (divides) the sender's data into smaller packets of data (maximum 1500 bytes). Each of these packets starts with an IP header that contains the destination address of the packet. The UDP system then adds a second header (the UDP control header) that includes a destination port. The packets are sent through the system where they may be received or lost in transmission. Because the UDP protocol does not contain any guarantee of delivery, it is up to the user on how to handle lost packets of data.

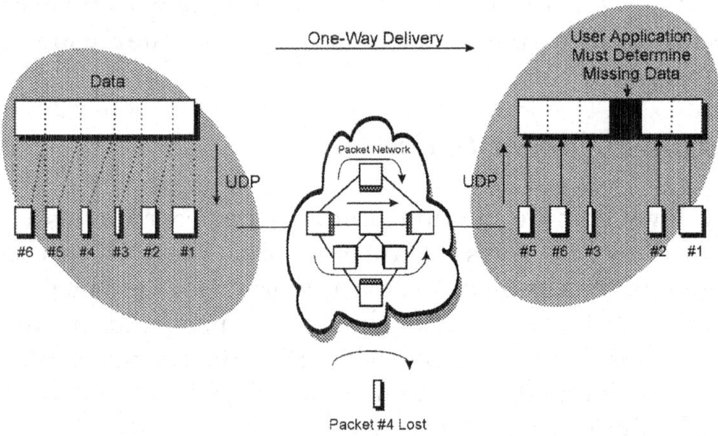

Figure 1.30, UDP Packet Transmission

Transmission Control Protocol (TCP)

Transmission control protocol is a session layer protocol that coordinates the transmission, reception, and retransmission of packets in a data network to ensure reliable (confirmed) communication. The TCP protocol coordinates the division of data information into packets, adds sequence and flow control information to the packets, and coordinates the confirmation and retransmission of packets that are lost during a communication session. TCP utilizes Internet Protocol (IP) as the network layer protocol.

Figure 1.31 shows how transaction control protocol (TCP) operates to reliably send data through a packet network. This diagram shows that the TCP system receives the data from a specific communication port (port number). The TCP system then packetizes (divides) the sender's data into smaller packets of data (maximum 1500 bytes). Each of these packets starts with an IP header that contains the destination address of the packet. The TCP system then adds a second header (the TCP control header) that includes a sequence number along with other flow control information. The packets are sent through the system where they may be received at different time periods. The sequence numbers can be used to reorder the packets. The TCP protocol also includes a window size that indicates to the receiving device how many packets it can receive before it must acknowledge their receipt.

This window defines how much data the sending device must keep in temporary memory to enable the retransmission of a packet in the event that a packet is lost in transmission. If a packet is lost, the receiving device requests the transmitting device to re-send the packet with a specific sequence number.

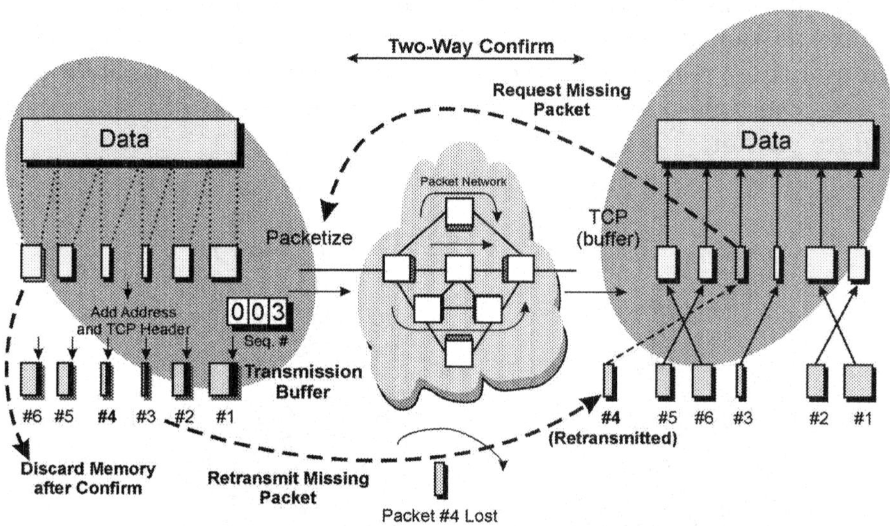

Figure 1.31, TCP Packet Transmission

Real Time Transport Protocol (RTP)

RTP is a packet based communication protocol that adds timing and sequence information to each packet to allow the reassembly of packets to reproduce real time audio and video information. RTP is defined in RFC 1889. RTP is the transport used in IP audio and video environments.

Figure 1.32 shows how real time transmission control protocol (RTP) operates to send real time data through a packet network that may have variable transmission delays. This diagram shows that an RTP system requires that a real time signal (e.g. audio signal) be converted to digital form (digital audio) prior to transmission. This digital signal is divided into small packets. The RTP protocol is a high-level protocol and each packet of data each of the transmitted packets starts with an IP header that contains the destination address of the packet. An additional flow control protocol header is added (usually UDP protocol header) to identify the specific port the data will be routed to at it's destination. The RTP system then adds a third header (the RTP control header). The RTP system uses a precise clock to add time stamp information to each packet along with other signal recreation control information. Because the packets may have different types of compression and their recreation time can dramatically vary, the RTP protocol header uses the time stamp and other information to decode the and recreate the data packet.

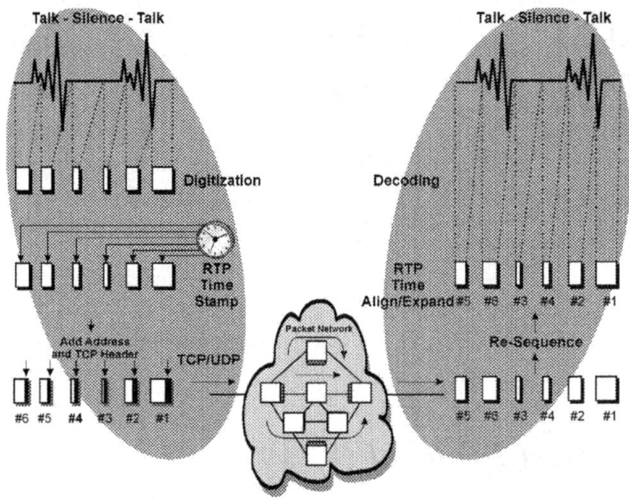

Figure 1.32, RTP Packet Transmission

Packet Buffering

Packet buffering is the process of using memory space to storing packets that are awaiting transmission or for storing a received packet. The memory may be located in the network interface controller or in the computer to which the controller is connected.

Figure 1.33 shows how packet buffering can be used to reduce the effects of packet delays and packet loss for streaming media systems. This diagram shows that during the transmission of packets from the media server to the viewer, some of the packet transmission time varies (jitter) and some of the packets are lost during transmission. The packet buffer temporarily stores data before providing it to the media player. This provides the time necessary to time synchronize the packets and to request and replace packets that have been lost during transmission.

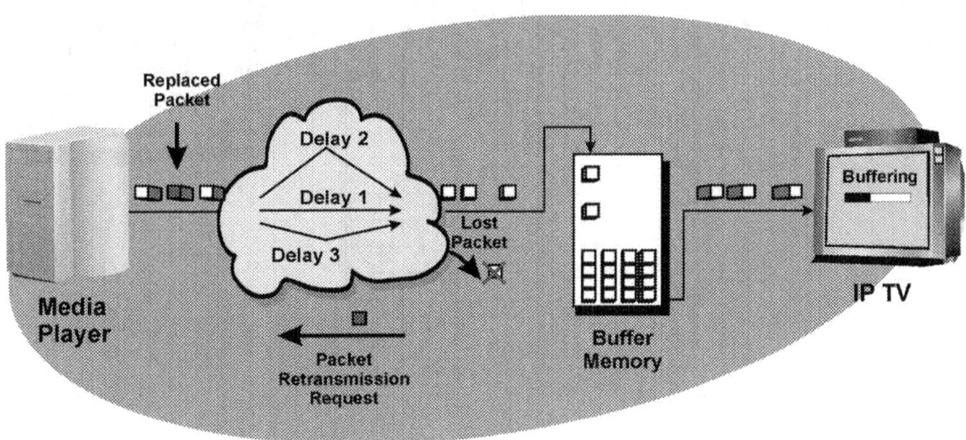

Figure 1.33, Packet Buffering

Packet buffering systems adds delay to the transmission of packets. This can be a significant challenge for real time systems such as video conferencing or IP television system channel changes. To overcome the delays introduced in packet buffering, data bursting may be used. Data bursting is the temporary increase in packet data transmission to rapidly fill up buffers or other packet queuing systems.

Figure 1.34 shows the use of high-speed data transmission (bursting data) at the beginning of a media streaming session can reduce the buffering time needed to start the playing of the video clip.

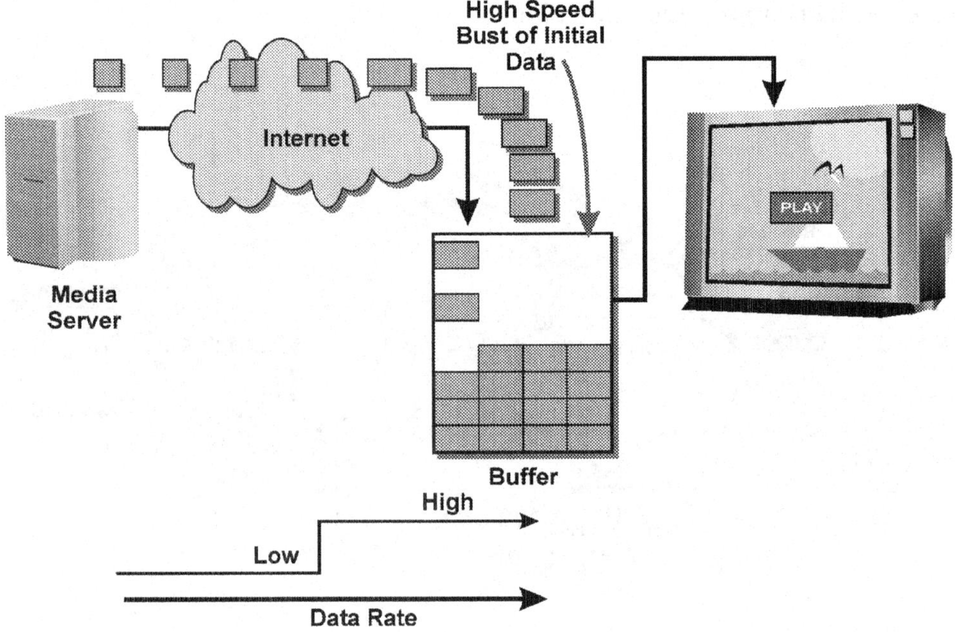

Figure 1.34, Rapid Video Play by Data Bursting

MPEG IP Packet Encapsulation

MPEG IP packet encapsulation is the process of inserting the entire contents of MPEG packets into the payload (data portion) of an IP packet. Several 188 byte MPEG packets can be encapsulated into a single IP packet. Each IP packet contains the IP header (address), UDP header (port), the RTP header (time) which is followed by several MPEG packets.

Bandwidth Control

Bandwidth control is the process of detecting and adjusting the bandwidth that is available or assigned to a particular device or service.

If the bandwidth available for a streaming media session is limited or it is decided to reduce or adjust the bandwidth for other reasons (e.g. high cost connection), rate shaping or stream thinning may be used to adjust the bandwidth that is assigned to a streaming connection service.

Rate shaping is the identification, categorization and prioritization of the transfer of data or information through a system or a network to match user requirements with network capacity and service capabilities.

Stream thinning is the process of removing some of the information in a media stream (such as removing image frames) to reduce the data transmission rate. Stream thinning may be used to reduce the quality of a media stream as an alternative to disconnecting the communication session due to bandwidth limitations.

Figure 1.35 shows how a media server can adjust its data transmission rate to compensate for different user data rates. This example shows how a media server is streaming packets to an end user (for an Internet television viewer). Some of the packets are lost at the receiving end of the connection because of the access device. The receiving device (a television set top box) sends back control packets to the media server indicating that the communication session is experiencing a higher than desirable packet or frame loss rate. The media server can use this information to change its media compression and data transmission rates to compensate for the slow user access link.

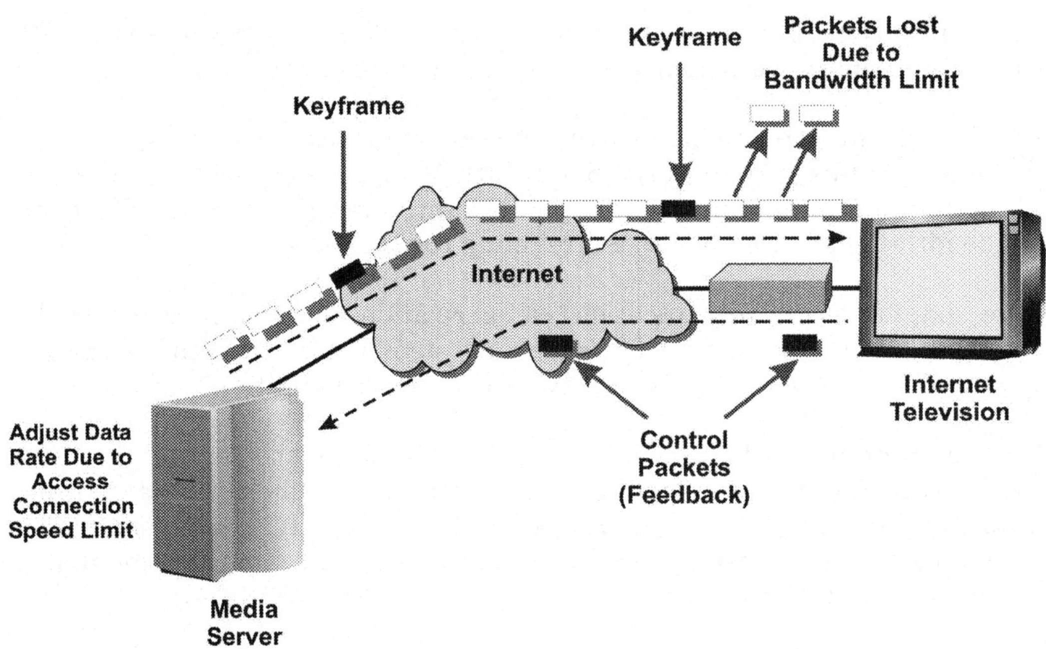

Figure 1.35, Digital Video Streaming Bandwidth Control

Protocol Oppression

Protocol oppression is the blocking or restricting of the transmission of data packets that contain specific types of protocols. Examples of protocol oppression is the blocking of UDP packets to restrict access to streaming media.

To overcome protocol oppression, some IP video providers may embed one protocol (such as UDP) into another protocol (such as TCP) which may be allowed to pass through firewalls.

Digital Video Quality (DVQ)

Digital video quality is the ability of a display or video transfer system to recreate the key characteristics of an original digital video signal. Digital video and transmission system impairments include tiling, error blocks, jerkiness, artifacts (edge busyness) and object retention.

Figure 1.36 shows some of the causes and effects of video distortion that may occur in IP Television systems. This example shows that video digitization and compression converts video into packets that can be sent through data networks (such as the Internet). Packet loss and packet corruption results in distorted video signals. This example shows that some types of distortion include tiling, error blocks and retained images.

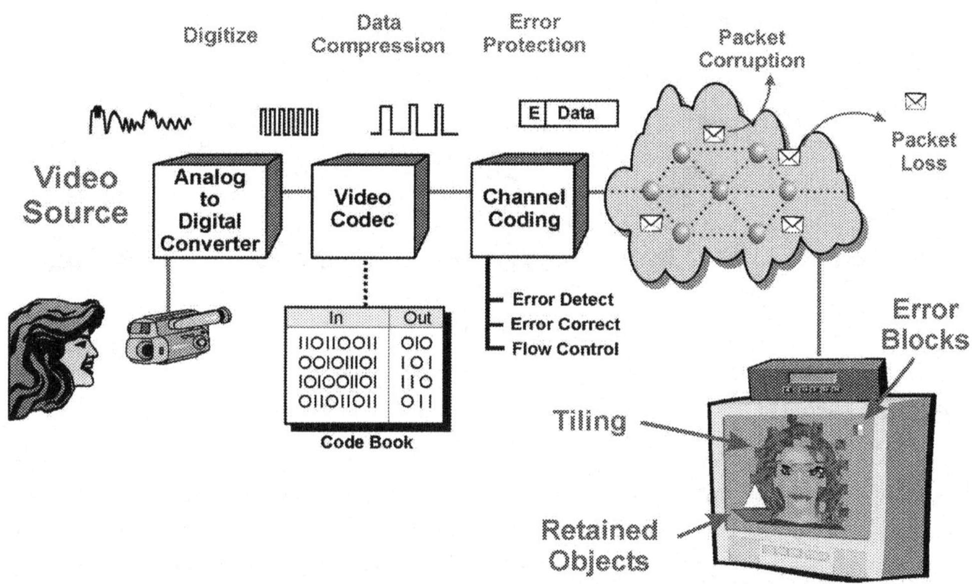

Figure 1.36, IP Video Distortion

Tiling

Tiling is the changing of a digital video image into square tiles that are located in positions other than their original positions on the screen.

Error Blocks

Error blocks are groups of image bits (a block of pixels) in a digital video signal that do not represent error signals rather than the original image bits that were supposed to be in that image block.

Figure 1.37 shows an example of how error blocks are displayed on a digital video signal. This diagram shows that transmission errors result in the loss of picture blocks. In this example, the error blocks continue to display until a new image that is received does not contain the errors.

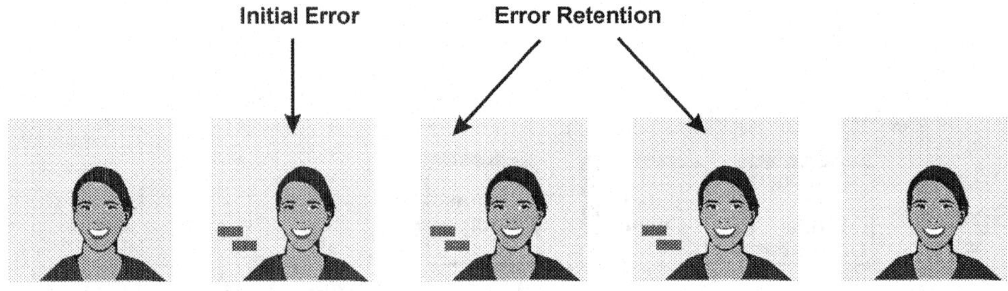

Figure 1.37, Digital Video Error Blocks

Jerkiness

Jerkiness is holding or skipping of video image frames or fields in a digital video. Jerkiness may occur when a significant number of burst errors occur during transmission that results in the inability of a receiver to display a new image frame. Instead, the digital video receiver may display the previous frame to minimize the perceived distortion (s jittery image is better than no image).

Aliasing Effects

Aliasing effects are unwanted distortions that result from the conversion of an image where the sampling of the image is at a speed less than half of the most rapid changes in the image. Aliasing effects commonly appear as lines or ripples in the scanned or converted image.

Artifacts

Artifacts, in general, are results, effects or modifications of the natural environment produced by people. In the processing or transmission of audio or video signals, a distortion or modification produced due to the actions of people or due to a process designed by people.

Mosquito noise is a blurring effect that occurs around the edges of image shapes that have a high contrast ratio. Mosquito noise can be created through the use of lossy compression when it is applied to objects that have sharp edges (such as text).

Figure 1.38 shows an example of mosquito noise artifacts. This diagram shows that the use of lossy compression on images that have sharp edges (such as text) can generate blurry images.

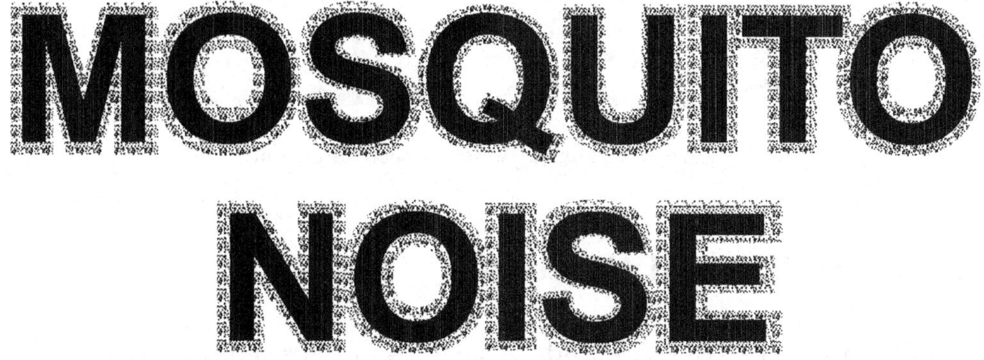

Figure 1.38, Mosquito Noise Artifacts

Object Retention

Object retention is the keeping of a portion of a frame or field on a digital video display when the image has changed. Object retention occurs when the data stream that represents the object becomes unusable the digital video receiver. The digital video receiver decides to keep displaying the existing object in successive frames until an error free frame can be received.

Figure 1.39 shows a how a compressed digital video signal may have objects retained when errors occur. This example shows that an original sequence where the images have been converted into objects. When the scene change occurs, some of the bits from image objects are received in error, which results in the objects remaining (a bird and the sail of a boat) in the next few images until an error free portion of the image is received.

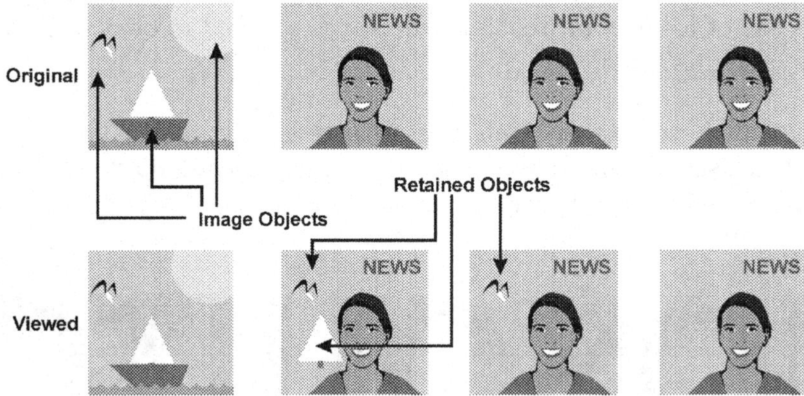

Figure 1.39, Digital Video Object Retention

Streaming Control Protocols

Streaming protocols are commands, processes and procedures that are used for delivering and controlling the real-time delivery of media (such as audio and or video streaming). Streaming control protocols control the setup, playing, pausing, recording and tear down of streaming sessions. Streaming control protocols that are used for IPTV systems include real time streaming protocol (RTSP) and digital storage media command and control (DSM-CC).

Real Time Streaming Protocol (RTSP)

Real time streaming protocol is an Internet protocol that is used for continuous (streaming) audio and video sessions. RTSP provides the control of playing, stopping, media position control (e.g. fast forward) via bi-directional (two-way) communication sessions. RTSP is defined in RFC 2326.

Digital Storage Media Command and Control (DSM-CC)

Digital storage media command and control is an MPEG extension that allows for the control of MPEG streaming. DSM-CC provides VCR type features.

Video Formats

Video formatting is the method that is used to contain or assign digital media within a file structure or media stream (data flow). Video formats are usually associated with specific standards like MPEG video format or software vendors like Quicktime MOV format or Windows Media WMA format.

Video formats can be a raw media file that is a collection of data (bits) that represents a flow of image information or it can be a container format that is a collection of data or media segments in one data file. A file container may hold the raw data files (e.g. digital audio and digital video) along with descriptive information (meta tags).

Some of the common video formats used in IPTV system include MPEG, Quicktime, Real media, Motion JPEG (MJPEG) and Windows Media (VC-1). Video formats that are used in the production or transfer process for video and television media include D1, D2, digital picture exchange (DPX), general exchange format (GXF), advanced authoring format (AAF) and material exchange format (MXF).

MPEG

Motion picture experts group develops digital video encoding standards. The MPEG standards specify the data compression and decompression processes and how they are delivered on digital broadcast systems.

The MPEG system defines the components (such as a media stream or channel) of a multimedia signal (such as a digital television channel) and how these channels are combined, transmitted, received, separated, synchronized and converted (rendered) back into a multimedia format. MPEG has several compression standards including MPEG-1, MPEG-2, MPEG-4 (original) and MPEG-4 AVC/H.264.

MPEG-1 offers less than standard television resolution. MPEG-1 was designed for slow speed stored media with moderate computer processing capabilities. MPEG-2 is designed and used for television broadcaster (radio, satellite and cable TV) of standard and high definition television. The MPEG-4 specification was designed for allows for television transmission over packet data networks such as broadband Internet. The initial release of the MPEG-4 system has the same amount of video compression as MPEG-2. The MPEG-4 system was enhanced with a 2nd type of compression called advanced video coding (AVC)/H.264 which increased the compression amount which added significant benefit to companies installing MPEG systems (more channels in less bandwidth).

Figure 1.40 shows the development timeline for MPEG compression. The first standard version of MPEG-1 was introduced in 1993 and was used for CD ROMs. Layer 3 of MPEG-1 provided the audio compression (called MP3). In 1995, MPEG-2 compression was released which has been used in digital television and satellite broadcasting system. In 1996, MPEG-3 was released. In 1999, MPEG-4 was released which efficiently provides digital video through packet data networks such as the Internet. In 2003, MPEG-4 was updated with advanced video coding (AVC) technology to provide improved video signals at higher compression rates (lower data transmission rates).

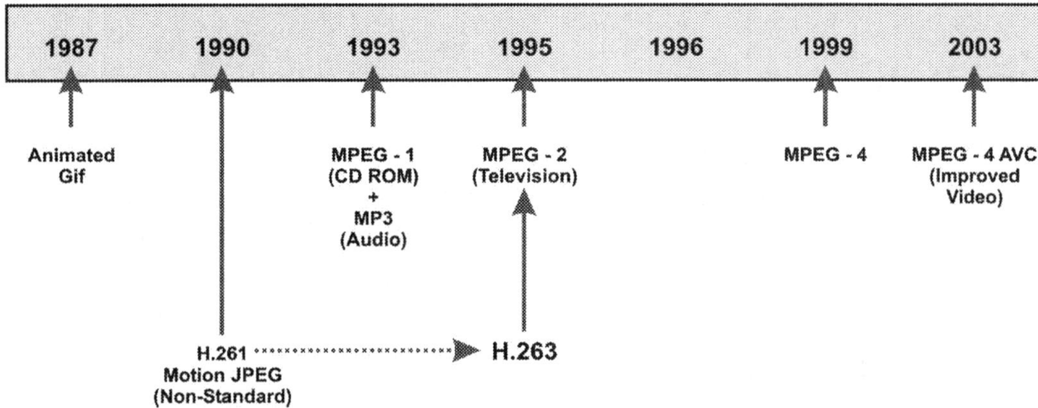

Figure 1.40, MPEG Codec Timeline

Figure 1.41 shows comparison of MPEG-2 and the new MPEG-4/AVC video coding system. This diagram shows that the standard MPEG-2 video compression system requires approximately 3.8 Mbps for standard definition (SD) television and 19 Mbps for high definition (HD) television. The MPEG-4 AVC video coding system requires approximately 1.8 Mbps for SD television and 6 to 8 Mbps for HD television.

	MPEG-2	MPEG-4/AVC or VC-1
Standard Definition (SD)	3.8 Mbps	1.8 Mbps
High Definition (HD)	19 Mbps	6-8 Mbps

Figure 1.41 MPEG Compression

Quicktime

Quicktime is a computer video format (sound, graphics and movies) that was developed by Apple computer in the early 1990s, QuickTime files are designated by the .mov extension.

Real Media

Real Media is a container format developed by the company "Real" and is used for streaming media files. Real Media files can use several types of coding processes. Real Media files are usually designated by the .rm file extension.

Motion JPEG (MJPEG)

A motion JPEG is a digital video format that is only composed of independent key frames. Because MJPEG does not use temporal compression, each video frame can be independently processed without referencing other frames.

Windows Media (VC-1)

Windows media .WM is a container file format that holds multiple types of media formats. The .WM file contains a header (beginning portion) that describes the types of media, their location and their characteristics that are contained within the media file. VC-1 is the designation for Microsoft's Windows Media Player codec by the SMPTE organization.

D1 Video Format

D1 is an uncompressed digital video format standard that is overseen by the Society of Motion Picture and Television Engineers (SMPTE). The D1 format is a component signal format which has a relatively high bandwidth requirement as compared to combined (composite) color formats.

D2 Video Format

D2 is an uncompressed digital video format standard that is overseen by the Society of Motion Picture and Television Engineers (SMPTE). The D2 format is a composite signal format which has a reduced bandwidth requirement as opposed to having data for each video component. The D2 format includes four audio channels and an analog cue channel.

Digital Picture Exchange (DPX)

Digital picture exchange is an uncompressed image file format. The DPX SMPTE 268M standard was approved in 1994.

General Exchange Format (GXF)

General exchange format is a broadcast media standard that allows for the sending of audio and video files in a time-multiplexed stream. GXF can send compressed audio and video media. The GXF standard is SMPTE 360M.

Advanced Authoring Format (AAF)

Advanced authoring format is a standard media container structure that is used to hold digital media (essence and metadata). In addition to holding the raw media in a format that can be used by production people, the AAF structure includes additional information that can assist in the production (authoring) process.

Material Exchange Format (MXF)

Material exchange format is a standard media container structure that is used to hold digital media (essence and metadata) that is used to transfer finished content. MXF is a limited version (subset) of the AAF structure. The metadata within the MXF media format describes how the media elements (e.g. video and audio) are re-assembled into their original media format.

Video Rendering

Video rendering is the process of converting media into a form that a human can view. An example of rendering is the conversion of a data file into a sequence of image that is displayed on video display.

Video Compositor

A video compositor is a device or system that can take two or more video inputs or graphic images and combine them into one composite video signal.

Figure 1.42 shows that a video compositor can take two or more video inputs or graphic images and combine them into one composite video signal. This example shows a video compositor that takes a video format (news clip), text graphics (scene caption) and a graphic image (explosion picture) and combines them (renders) onto a single video display.

Spatial Scalability

Spatial scalability is the ability of a media file or picture image to reduce or vary the number of image components or data elements representing that that picture over a given area (spatial area) without significantly changing the quality or resolution of the image.

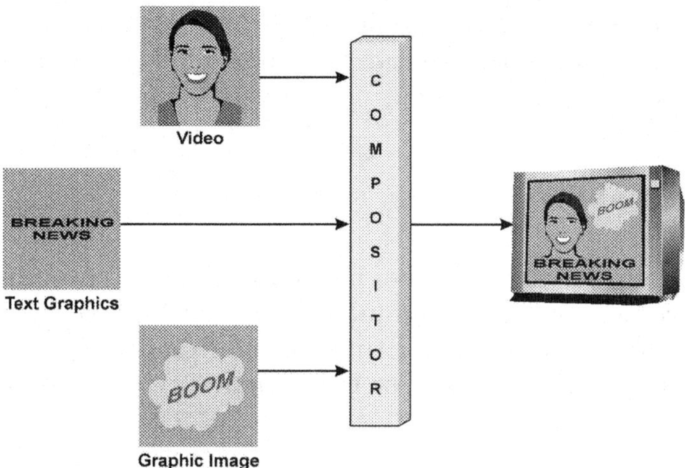

Figure 1.42, Video Compositor

Figure 1.43 shows how spatial scalability can be used to provide images in different formats through the use of progressive adding of image information. This example shows that a small portion of a data signal (a low data rate) is used to provide a low resolution mobile video (MV). Additional bits

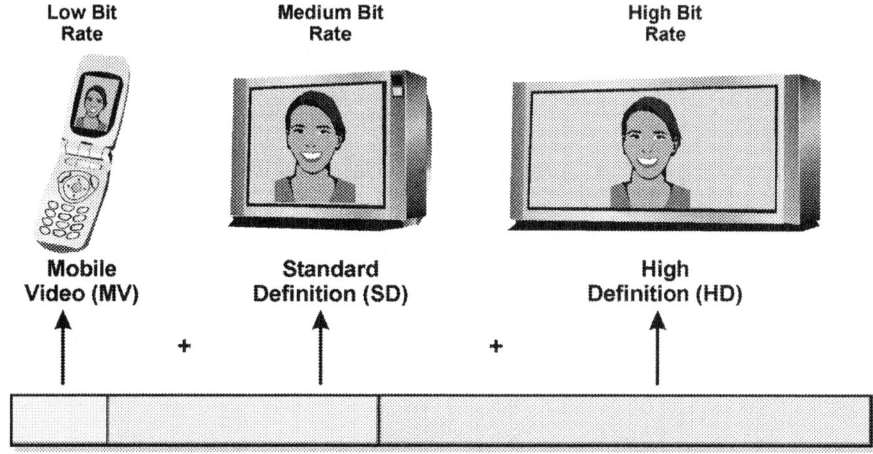

Figure 1.43, Spatial Scalability

are added to the low resolution image data produce a standard definition (SD) video image. Additional bits can then be added to produce a high definition (HD) video image.

Temporal Scalability

Temporal scalability is the ability of a streaming media program or moving picture file to reduce or vary the number image or data elements representing that media file for a particular time period (temporal segment) without significantly changing the quality or resolution of the media over time.

Loss Concealment

Loss concealment is a process that attempts to alter a signal or display in a way that reduces the effects of errors or packet losses. An example of a loss concealment process is the insertion of a previous frame in a video sequence that has had a fame lost.

Appendix 1

Acronyms

50i-50 Interlaced
50p-50 Progressive
60i-60 Interlaced
60p-60 Progressive
AAF-Advanced Authoring Format
ASF-Advanced Streaming Format
ASI-Asynchronous Serial Interface
ASX-Advanced Streaming Index
AVC-Advanced Video Coding
AVI-Audio Video Interleaved
B-Frame-Bidirectional Frame
BIFF-Binary Interchange File Format
BMP-BitMaP
CC-Closed Caption
CIF-Common Intermediate Format
CSR-Connection Success Rate
CVBS-Composite Video Baseband Signal
DAS-Digital Audio Server
DCT-Discrete Cosine Transform
DFT-Discrete Fourier Tranform
DPI-Digital Program Insertion
dpi-Dots Per Inch
DPX-Digital Picture eXchange
DR-Dynamic Range
DSM-CC-Digital Storage Media Command and Control
DTLA-Digital Transmission Licensing Administrator
DTS-Digital Theater Sound
DV Camcorder-Digital Video Camcorder
DVB-Digital Video Broadcast
DVB-H-Digital Video Broadcasting Handheld
DVB-T-Digital Video Broadcasting Terrestrial
DVE-Digital Video Effect
DVI-Digital Video Interactive
DVQ-Digital Video Quality
fps-Frames Per Second
GIF-Graphics Interchange Format
GOP-Group of Pictures
GXF-General eXchange Format
HD-High Definition
HDMI-High Definition Multimedia Interface
HTTP-Hypertext Transfer Protocol
IDCT-Inverse Discrete Cosine Transform
I-Frame-Intra Frame
IP-Internet Protocol
ISO-International Standards Organization
JPEG-Joint Photographic Experts Group
KQI-Key Quality Indicators
LPI-Lines Per Inch
MJPEG-Motion JPEG
MP3-Motion Picture Experts Group Layer 3
MP4-MPEG-4
MPEG-Motion Picture Experts Group
MV-Mobile Video
MXF-Material eXchange Format
NSV-Nullsoft Video
NTSC-National Television System Committee

OSD-On Screen Display
PAL-Phase Alternating Line
PDV-Packet Delay Variation
P-Frame-Predicted Frame
PIG-Picture in Graphics
PIP-Picture in Picture
Play-List-Playlist
PMS-Panatone Matching System
PSNR-Peak Signal to Noise Ratio
QCIF-Quarter Common Interchange Format
QCIF-Quarter Common Intermediate Format
QoS Policy-Quality of Service Policy
QoS-Quality Of Service
QT Atoms-Quicktime Atoms
RGB-Red, Green, Blue
RLC-Run Length Coding
RLE-Run Length Encoding
RM-RealMedia
RTN-Real Time Notification
RTP-Real Time Transport Protocol
RTSP-Real Time Streaming Protocol
SD-Standard Definition
SIF-Source Intermediate Format
SMPTE-Society of Motion Picture and Television Engineers
S-Video-Separate Video
SWF-Shockwave Flash
Sync Impairment-Synchronization Impairments
TCP-Transmission Control Protocol
TV Portal-Television Portal
UDP-User Datagram Protocol
VBI-Video Blanking Interval
VBV-Video Buffer Verifier
VGA-Video Graphics Adapter
VLC-Variable Length Coding
VLE-Variable Length Encoding
VQM-Video Quality Measurement
VRML-Virtual Reality Modeling Language
WM-Windows Media
Y-Luminance Signal

Index

Althos Publishing Book List

Product ID	Title	# Pages	ISBN	Price	Copyright
Billing					
BK7727874	Introduction to Telecom Billing	48	0974278742	$11.99	2004
BK7769438	Introduction to Wireless Billing	44	097469438X	$14.99	2004
Business					
BK7781359	How to Get Private Business Loans	56	1932813594	$14.99	2005
BK7781368	Career Coach	92	1932813683	$14.99	2006
Datacom					
BK7727873	Introduction to Data Networks	48	0974278734	$11.99	2003
IP Telephony					
BK7727877	Introduction to IP Telephony	80	0974278777	$12.99	2003
BK7781361	Tehrani's IP Telephony Dictionary, 2nd Edition	628	1932813616	$39.99	2005
BK7780530	Internet Telephone Basics	224	0972805303	$29.99	2003
BK7780532	Voice over Data Networks for Managers	348	097280532X	$49.99	2003
BK7780538	Introduction to SIP IP Telephony Systems	144	0972805389	$14.99	2003
BK7781311	Creating RFPs for IP Telephony Communication Systems	86	193281311X	$19.99	2004
BK7781309	IP Telephony Basics	324	1932813098	$34.99	2004
BK7769430	Introduction to SS7 and IP	56	0974694304	$12.99	2004
IP Television					
BK7781362	Creating RFPs for IP Television Systems	86	1932813624	$19.99	2005
BK7781357	IP Television Directory	154	1932813578	$89.99	2005
BK7781355	Introduction to Data Multicasting	68	1932813551	$14.99	2005
BK7781340	Introduction to Digital Rights Management (DRM)	84	1932813403	$14.99	2005
BK7781351	Introduction to IP Audio	64	1932813519	$14.99	2005
BK7781335	Introduction to IP Television	104	1932813357	$14.99	2005
BK7781330	Introduction to IP Video Servers	68	1932813306	$14.99	2005
BK7781341	Introduction to IP Video	88	1932813411	$14.99	2005
BK7781352	Introduction to Mobile Video	68	1932813527	$14.99	2005
BK7781353	Introduction to MPEG	72	1932813535	$14.99	2005
BK7781342	Introduction to Premises Distribution Networks (PDN)	68	193281342X	$14.99	2005
BK7781354	Introduction to Telephone Company Television (Telco TV)	84	1932813543	$14.99	2005
BK7781344	Introduction to Video on Demand (VOD)	68	1932813446	$14.99	2005
BK7781356	IP Television Basics	308	193281356X	$34.99	2005
BK7781334	IP TV Dictionary	652	1932813349	$39.99	2005
BK7781363	IP Video Basics	280	1932813632	$34.99	2005
Programming					
BK7727875	Wireless Markup Language (WML)	287	0974278750	$34.99	2003
BK7781300	Introduction to xHTML:	58	1932813004	$14.99	2004
Legal and Regulatory					
BK7769433	Practical Patent Strategies Used by Successful Companies	48	0974694339	$14.99	2003
BK7781332	Strategic Patent Planning for Software Companies	58	1932813322	$14.99	2004
BK7780533	Patent or Perish	220	0972805338	$39.95	2003
Telecom					
BK7727872	Introduction to Private Telephone Systems 2nd Edition	86	0974278726	$14.99	2005
BK7727876	Introduction to Public Switched Telephone 2nd Edition	54	0974278769	$14.99	2005
BK7780537	SS7 Basics, 3rd Edition	276	0972805370	$34.99	2003
BK7780535	Telecom Basics, 3rd Edition	354	0972805354	$29.99	2003
BK7727870	Introduction to Transmission Systems	52	097427870X	$14.99	2004
BK7781313	ATM Basics	156	1932813136	$29.99	2004
BK7781302	Introduction to SS7	138	1932813020	$19.99	2004
BK7781345	Introduction to Digital Subscriber Line (DSL)	72	1932813454	$14.99	2005

For a complete list please visit
www.AlthosBooks.com

Althos Publishing Book List

Winter 2005-06

Wireless					
BK7781306	Introduction to GPRS and EDGE	98	1932813063	$14.99	2004
BK7781304	Introduction to GSM	110	1932813047	$14.99	2004
BK7727878	Introduction to Satellite Systems	72	0974278785	$14.99	2005
BK7727879	Introduction to Wireless Systems	536	0974278793	$11.99	2003
BK7769432	Introduction to Mobile Telephone Systems	48	0974694320	$10.99	2003
BK7769435	Introduction to Bluetooth	60	0974694355	$14.99	2004
BK7769436	Introduction to Private Land Mobile Radio	50	0974694363	$14.99	2004
BK7769434	Introduction to 802.11 Wireless LAN (WLAN)	62	0974694347	$14.99	2004
BK7769437	Introduction to Paging Systems	42	0974694371	$14.99	2004
BK7781308	Introduction to EVDO	84	193281308X	$14.99	2004
BK7781305	Introduction to Code Division Multiple Access (CDMA)	100	1932813055	$14.99	2004
BK7781303	Wireless Technology Basics	50	1932813039	$12.99	2004
BK7781312	Introduction to WCDMA	112	1932813128	$14.99	2004
BK7780534	Wireless Systems	536	0972805346	$34.99	2004
BK7769431	Wireless Dictionary	670	0974694312	$39.99	2005
BK7769439	Introduction to Mobile Data	62	0974694398	$14.99	2005
Optical					
BK7781329	Introduction to Optical Communication	132	1932813292	$14.99	2006

Order Form

Phone: 1 919-557-2260
Fax: 1 919-557-2261 **Date:**______________
404 Wake Chapel Rd., Fuquay-Varina, NC 27526 USA

Name:______________________________ Title:______________________________
Company:____
Shipping Address:____
City:____________________________ State:______________ Postal/ Zip:______________
Billing Address:____
City:____________________________ State:______________ Postal/ Zip ______________
Telephone:____________________________ Fax:______________________________
Email: ____
Payment (select): VISA ___ AMEX ___ MC ___ Check ___
Credit Card #: ______________________________Expiration Date: ______________
Exact Name on Card: ______________________________

Qty.	Product ID	ISBN	Title	Price Ea	Total
Book Total:					
Sales Tax (North Carolina Residents please add 7% sales tax)					
Shipping: $5 per book in the USA, $10 per book outside USA (most countries). Lower shipping and rates may be available online.					
Total order:					

For a complete list please visit
www.AlthosBooks.com

CPSIA information can be obtained
at www.ICGtesting.com
Printed in the USA
LVOW03s0827210316
480054LV00008B/321/P